D1357751

MANNA: AN HISTORICAL GEOGRAPHY

BIOGEOGRAPHICA

Editor-in-Chief

J. SCHMITHÜSEN

Editorial Board

L. BRUNDIN, Stockholm; H. ELLENBERG, Göttingen; J. ILLIES, Schlitz;
H. J. JUSATZ, Heidelberg; C. KOSSWIG, Istanbul; A. W. KÜCHLER, Lawrence;
H. LAMPRECHT, Göttingen; A. MIYAWAKI, Yokohama; W. F. REINIG, Hardt;
S. RUFFO, Verona; H. SICK, Rio de Janeiro; H. SIOLI, Plön;
V. VARESCHI, Caracas; E. M. YATES, London

Secretary

P. MÜLLER, Saarbrücken

VOLUME 17

DR. W. JUNK B.V. PUBLISHERS
THE HAGUE-BOSTON-LONDON 1980

MANNA: AN HISTORICAL GEOGRAPHY

by

R. A. DONKIN

(Fellow of Jesus College, Cambridge)

UNIVERSITY LIBRARY
19 JAN 1981
LANCASTER

DR. W. JUNK B.V. PUBLISHERS
THE HAGUE-BOSTON-LONDON 1980

Distributors:

for the United States and Canada

Kluwer Boston, Inc.
160 Old Derby Street
Hingham, MA 02043
USA

for all other countries

Kluwer Academic Publishers Group
Distribution Center
P.O. Box 322
3300 AH Dordrecht
The Netherlands

Library of Congress Cataloging in Publication Data

Donkin, R   A
   Manna, an historical geography.

   (Biogeographica: v. 17)
   Bibliography
   Includes index.
   1. Manna plants. 2. Manna plants —Geographical distribution. 3. Manna. 4. Honeydew. I.
Title. II. Series: Biogeographica (Hague); v.
SB317.M33D66   633'.8   79–24368

ISBN 90 6193 218 1 (this volume)
ISBN 90 6193 884 8 (series)

8O OO4688

Copyright © 1980 Dr. W. Junk bv Publishers, The Hague.

All rights reserved. No part of this publication may be reproduced, stored in a retrieval system, or transmitted in any form or by any means, mechanical, photocopying, recording, or otherwise, without the prior written permission of the publishers Dr. W. Junk bv Publishers, P.O. Box 13713, 2501 ES The Hague, The Netherlands.

PRINTED IN THE NETHERLANDS

CONTENTS

Map 1. Approximate distribution of sugar cane, *Saccharum officinarum* L. (A: before the beginning of the Christian Era; B: expansion by A.D. 1500, after Bertin *et. al.*, 1971) and areas of collection of some principal mannas (1: *tar-angubīn*, on *Alhagi* spp.; 2: *schir-khecht* on *Cotoneaster nummularia* and *Atraphaxis spinosa*; 3: *gaz-ālāfi* on *Quercus* spp.; 4: *gaz-angubīn* on *Tamarix* spp.).

Map 2. A: Combined distribution of *Alhagi maurorum* Desv., *A. camelorum* Fisch., and *A. graecorum* Boiss. B: Collection of alhagi manna.

Map 3. Places named in references to the collection of alhagi manna.

Map 4. Approximate distribution of *Echinops persicus* Stev. et Fisch. (A–F) and of *Calotropis procera* R. Br. (1–20).

Map 5. Approximate distribution of *Atraphaxis spinosa* L. (A–R) and of *Cotoneaster nummularia* Fisch. et Meyer (1–18).

Map 6. Approximate limits of areas producing and marketing *schir-khecht* (on *Atraphaxis spinosa* L. and *Cotoneaster nummularia* Fisch. et Meyer).

Map 7. *Lecanora esculenta* Evers., western and central Asia.

Map 8. *Lecanora esculenta* Evers., North Africa.

Map 9. Combined distribution of *Quercus persica* Jaub. et Spach, *Q. brantii* Lindl., and *Q. mannifera* Lindl.

Map 10. Principal area of collection of willow manna, *bīd-angubīn* or *bīd-khecht*.

Map 11. Tamarisk manna reported, 1–4 on *Tamarix gallica* (*mannifera*) Ehren., 5 on *T. dioica* Roxb., *T. pentandra* Pall., and *T. aphylla* (*articulata*) Karst.

Map 12. Peninsula of Sinai.

Map 13. Combined distribution of *Astragalus adscendens* Boiss. et Haussk., *A. florulentus* Boiss. et Haussk., and *A. fasciculaefolius* Boiss.

Map 14. Approximate distributions of *Fraxinus ornus* L., *F. floribunda* Wall., and *F. rotundifolia* Mill.

Map 15. Manna of *Fraxinus* spp. in central and southern Italy and Sicily. Places mentioned in 16th- to 19th-century sources.

Map. 16. Approximate areas of collection of "Polish manna" (*Glyceria fluitans* R. Br.); 16th- to 19th-century sources.

Map 17. Regions and places mentioned in connection with the mannas of Australia.

Fig. 1. "The Gathering of Manna" Tintoretto, ca. 1594 (San Giorgio Maggiore, Venice).

Fig. 2. Hieroglyph for "manna" (fourth row from bottom, far left), Ptolemaic temple of Horus, Edfu (Dümichen, 1867: 1: Taf. LXXXIII).

Fig. 3. *Hedysarum alhagi* L. (Sibthorp, 1806–1840: 8: tab. 720); *Alhagi maurorum* Medic. (1787) Desv. (1813).

Fig. 4. *Calotropis procera* R. Br. (Alpinus, 1592: pp. 36–37).

Fig. 5. *Atraphaxis spinosa* L. (Dillenius, 1732: 1: tab. XL).

Fig. 6. *Cotoneaster nummularia* Fisch. et Meyer (Aitchison, 1888–1894: pl. 9).

Fig. 7. *Echinops persicus* Stev. et Fisch. ex Fisch. (Moghadam, 1930: pl. II).

Fig. 8. *Tamarix* (*gallica*) *mannifera* Ehren. (Ehrenberg and Hemprich [1820–1825], 1930: Taf. II).

Fig. 9. *Tamarix articulata* Vahl (Vahl, 1790–1794: 2: tab. XXXII).

Fig. 10. "Rain of Manna" Inscription, Sinai (P. Thomae Obecini Novariensis, in Kircher, 1652–1654: 2: p. 120).

Fig. 11. Account of the manna of Sinai. Bernhard von Breydenbach, 1483–1484 (1502: no pagination).

Fig. 12. Collecting manna, Sicily (Houel, 1782–1787: 1: pl. 32).

Fig. 13. Manna-producing species (Pomet, 1694: 1: p. 236): *Fraxinus ornus* L. (Manne de Calabre), *Alhagi* sp. (Manne Liquide), *Larix europaea* D. C. (Manne de Briançon).

Fig. 14. *Glyceria fluitans* R. Br. (Matthiolus, 1565: p. 1000).

Fig. 15. *Glyceria fluitans* R. Br. (Dalechamps, 1586–1587: 1: p. 414).

Fig. 16. *Larix europaea* D.C. [*Pinus larix* L.] (Lobel [1570–1571], 1576: pt. 2: p. 24).

Fig. 1. "The Gathering of Manna" Tintoretto, ca. 1594 (San Giorgio Maggiore, Venice).

1 INTRODUCTION

> And lucent syrops, tinct with cinnamon;
> Manna and dates, in argosy transferr'd
> From Fez; and spiced dainties, every one,
> From silken Samarcand to cedar'd Lebanon.
>
> John Keats *The Eve of St. Agnes*

The description "manna" has no uniform or precise meaning. In nature, it is chiefly, but not exclusively, applied to two composite categories of saccharine substances: (a) exudations from the branches or leaves of plants or trees, occasioned by unusually high atmospheric temperatures, or by the punctures of insects or artificial incisions, and (b) excretions of insects, either in the form of honeydew or, exceptionally, of protective cocoons. In the first category, the part played by insects is still imperfectly understood and some "exudations" may, on closer examination, turn out to be excretions. This appears to have been demonstrated in the case of manna found on the branches of species of *Tamarix*. Honeydew is produced by insects belonging to the order Rhynchota (many Coccina, all the Aleyrodina and Pysllina, and most of the Aphina and Cicadina); the description "manna" may be applied where the droplets solidify under conditions of low humidity and high atmospheric temperature. Manna of all kinds is chiefly associated with the hot, dry lands of the Old World. It has been less frequently reported from the Americas.

The nature of manna was apparently understood by some Arab and Persian scholars of the medieval period. But among others and in the West generally until early modern times, the substance was thought to be, like honey, a kind of "dew," the one collected by bees, the other by man[1] (honeydew and manna are also collected by bees when nectar is in short supply). From the late 16th century, European residents or travellers in the East came to accept the better informed Arab view. "All *mana*," affirmed Pedro Teixeira (ca. 1590), "is gum produced by one tree or another, like other gums, and the stories of its coming of dew are inventions or based on bad evidence."[2] Other "mannas," however, are quite different products. The most famous is the lichen *Lecanora esculenta* Evers., distributed by the wind in central and western Asia and collected chiefly in times of famine. The term has also been applied to certain fungi and even to a number of wild grains, notably *Glyceria fluitans* R. Br. ("Polish manna") with a slightly sweet taste and formerly regarded as a luxury. Manna was never an ordinary

---

[1] Al-Bīrūnī (973–1048), 1973: 1: p. 309, quoting Yaḥyā ibn Sarāfyūn ("Serapion the Elder," 9th century). See also Vincentius (died ca. 1260), 1494: pp. 47 (*manna*), 48 (*mel*); F. Fabri (1483), 1892–1893: 2: p. 545; Leclerc in 'Abd ar-Razzāq, 1874: pp. 342–343 (item 876). Miguel de Castanhoso ([*Abyssinia*, 1541–1543] 1902: p. 59) equated manna with honey. The Chinese of the 12th and 13th centuries, who knew manna only at second hand, used the term *kan-lu*, "sweet dew" (Hirth and Rookhill [Chu-fan-chi], 1911: p. 140 n. 1; Chavannes and Pelliot, 1913: p. 131 n. 3; Laufer, 1919: p. 345 n. 2; Stuart, 1928: p. 258).

[2] Teixeira, 1902: p. 204.

or regular item of diet, but rather something unusual and adventitious, a bonus or "gift." Several kinds were valued largely or exclusively as medicinal products. This doubtless contributed to the air of mystery which surrounded their mode of origin.

Out of the tradition of biblical manna, the name came to be applied to a variety of mythological or miraculous substances. In the early Middle Ages, manna was described as "angels' food"[3] and, from the 14th century at least, it was sometimes associated with the tombs or bodies of saints.[4] The word is also often used symbolically, in the sense of nourishment, including spiritual nourishment, *manna animae*.[5] There may be some notional connection with ambrosia and "nectar of the gods," the Sanskrit *amṛta* or *amṛita*, and with other magical foods. The Japanese Buddhist monk Ennin (Jikaku Daishi) of the early 9th century is said to have been restored to health after "a marvellous dream in which he received a honey-like medicine from heaven."[6]

Sweet, edible substances have probably always appealed to man and, until modern times, they have usually been in short or intermittent supply. Few sources of sugar have remained unexploited and most were probably recognised before the advent of agriculture. The aboriginal peoples of Australia extracted nectar from flowers[7] and collected the "sugar ant," *Melophorus* [*Camponotus*] *inflatus*,[8] as well as several kinds of manna. Saccharine products have been recommended as medicine (honey and the earliest granulated sugar, in addition to manna), added to alcoholic beverages, valued as preservatives, and of course made into a wide variety of sweetmeats. A craving for sugar appears to be particularly common among peoples of the Old World deserts and steppes.[9] For this there may be some physiological explanation. Persia and Turkestan became famous for their sweetmeats and it is here too that mannas were most fully appreciated by both sedentary and nomadic folk. West-central Asia received sugar cane, *Saccharum officinarum* L. (Map 1), at a relatively late date (ca. 6th century), but then made important advances in refining. Before this, honey, several mannas and honeydews, grape juice and must[10] and, in the south, date

---

[3] Baxter and Johnson, 1934 ("manna"). Cf. Mendeville (ca. 1350), 1953: 1: pp. 61, 109; 2: pp. 274, 449.

[4] Niccolò of Poggibonsi (1349), 1945: p. 107, and Fescobaldi (1384), 1948: p. 58, both concerning the body of St. Catherine, at the monastery of the same name in Sinai. John Mendeville ([ca. 1350] 1953: 1: p. 16) speculated that "In the tomb of St John [the Evangelist, in Ephesus] men may find nothing but manna, for some say his body was translated into Paradise." Henry Swinburne (1783–1785: 1: p. 194) observed that "miraculous water" from beneath the altar of the priory of St. Nicholas in Bari was known as "manna."

[5] Segneri (1624–1694), 1879.

[6] Reischauer, 1955: p. 22.

[7] Johnston and Cleland, 1942: p. 99; Irving, 1957: p. 122. Also reported (Ford, 1976: p. 320) from the New World.

[8] Mountford, 1945: pp. 158–159.

[9] Murav'ev, 1871: p. 161; Bodenheimer, 1947: p. 3.

[10] As late as ca. 1840, Josiah Harlan (1939: p. 49) found that in the province of Bulkh (Bactria) grapes were chiefly used to prepare syrup, *sheerah*, "sold at half the price of coarse dark-coloured sugar imported from the south."

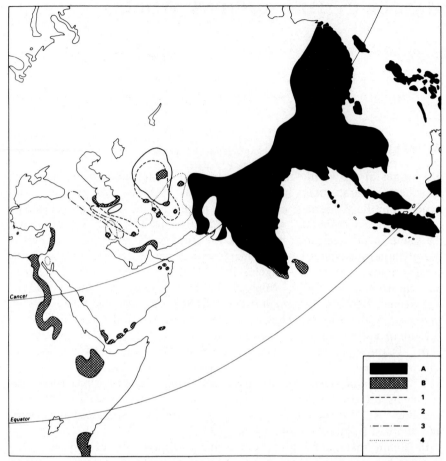

Map. 1. Approximate distribution of sugar cane, *Saccharum officinarum* L. (A: before the beginning of the Christian Era; B: expansion by A.D. 1500, after Bertin *et al.*, 1971) and areas of collection of some principal mannas (1: *tar-angubīn*, on *Alhagi* spp.; 2: *schir-khecht* on *Cotoneaster nummularia* and *Atraphaxis spinosa*; 3: *gaz-ālāfi* on *Quercus* spp.; 4: *gaz-agubīn* on *Tamarix* spp.).

syrup,[11] provided the chief forms of sweetening. The collection and use of manna fall within an ancient and well defined pattern of interest in sugar for food and medicine, the two uses overlapping. The spread and ultimate dominance of cane sugar have not entirely displaced the use of manna (and other sweeteners) for the properties differ, and local recipes and traditional remedies have preserved the preferences and knowledge of past centuries.

---

[11] Ibn al-Balkhī (ca. 1100), 1912: pp. 312 (dates and raisin syrup, *dūshāb*), 331 (honey). For early (cuneiform) evidence of the date palm and its products, see Landsberger, 1967.

3

2 MANNA IN THE ANCIENT WORLD

A. "MANNA"

The origin of the word "manna" has not been satisfactorily explained. It may have several roots, including the early Hebrew *mân* (what?). The Israelites in the Wilderness of Sin, seeing manna for the first time, are said to have exclaimed *mân-hû*, "what is this?" (Exodus 16: 15).[1] Subsequently, the name of the substance itself took the form of the interrogative. This is the etymology advanced by Flavius Josephus (ca. A.D. 94)[2] and by later commentators such as Fr. Angelus Palea (1550),[3] Johann Buxtorf (*Dissertatio de Manna*, ca. 1600),[4] Michael Walther (*Tractatu de Mannâ*, 1633),[5] and Samuel Bochart (*Geographia Sacra*, 1692).[6] *Mân* passed into Egyptian (*mennu*), Arabic (*mann*), Hellenistic Greek (μάννα) and Latin (*manna*).[7]

Modern authorities have pointed out that the Arabic *mann* also means "gift," in the sense of "free gift,"[8] "gift from God" or "gift from heaven" (*mann as-samā*).[9] The Jewish physician Maimonides (Moses ben Maimon, 1135–1204), who worked in Cairo, gives *mann* and *rizq* ("provision") as synonyms for the Persian manna *tar-angubīn*.[10] It is possible that *mân*, meaning "gift" or something similar, was originally a Sinaitic dialect word and adopted by the Hebrews for manna.

P. Haupt maintained that "the primary connotation of Hebrew *mân* [related to *min*, "from"] is .... separation, elimination, secretion," which could refer to the mode of origin of manna, and further that "*mân-hû* [Syriac *mânâ-hû*] is Aramaic, not Hebrew .... the popular etymology given in Exodus 16: 15 must be a late gloss."[11]

---

[1] Commentary by Zenner, 1899: pp. 164–166.
[2] Josephus, 1967–1969: 1: pp. 331–335.
[3] Palea, 1550: p. 251.
[4] Buxtorf, 1747: cols. 587–592 (*De Nomine Mannae*).
[5] M. Walther, 1633: pp. 16–17.
[6] Bochart, 1692: 3: p. 59.
[7] Watt (1889–1893: 5: p. 165) gives similar forms (*ména, manná*) from India and Malaya. Cf. Portuguese and Spanish *maná* (Meyerhof, 1938: p. 7).
[8] Stillé, 1868: 2: p. 438.
[9] Guest and Townsend, 1966–1974: 3: p. 502. C. R. Markham (ed. Orta, 1913: p. 281 n.) observed: "[Manna] is still known throughout India, as throughout Europe, by its Egyptian and Hebrew and Arabic names [*mēna* of the Malabar coast, *mānā* of Hindustan and the Deccan], meaning "a gift given us." Charles Thompson (1744: 3: p. 199) noted Hebrew *manah*, "a gift." See also Lengerke, 1844: p. 444; Kolb, 1892: p. 1; Haupt, 1922: pp. 235–236; Kaiser, 1924: pp. 100–102; Anon., 1970: p. 108.
[10] Maimonides, 1940: p. 193 (item 386).
[11] Haupt, 1922: pp. 235–236. Dorvault (1884: p. 301) derived *manne* (French) from *manare*, "to flow" or "to ooze."

It has also been suggested that the origin of "manna" may be found in Sumero-Akkadian mythology. "The tree sacred to Anu [sky or heaven god] was called *ma-nu* in Sumerian, and is persistently connected with the tamarisk and date palm in the texts. Not impossibly is the Hebrew term taken directly from this Sumerian word."[12] Manna was generally thought to descend from the sky, and the tamarisk was the most likely natural source of Sinaitic manna (*infra* pp. 72–79).

B. MESOPOTAMIA

R. Campbell Thompson (*Dictionary of Assyrian Botany*, 1949) gave extended attention to words in Sumerian, Akkadian and Assyrian connected with manna. They seem to show that several different products were in use at a very early date and that their modes of origin were more fully understood than during some later periods.

Persian *qudrat ḥalwā* and Turkish *küdret helvasi* ("potent sweetmeat," "manna," figuratively "divine sweetmeat," "manna of the Israelites")[13] appear to come from the Assyrian *qudru*.[14] The equivalent Sumerian word meant "earth of the storm-[wind-, rain-] god." It is possible that this was not one of the exuded or excreted mannas, but rather the lichen *Lecanora esculenta* (not considered by Thompson). Deposits of *L. esculenta* have usually been found after spells of high wind and generally stormy weather (*infra*, p. 49).

The most likely identification of the Assyrian *supalu* (Sumerian, "earth of the moon-god") is the oak manna of Kurdistan (*infra*, p. 56).[15] Chiefly remarkable perhaps is Thompson's argument that the Assyrian word *qaqqadânu* refers to the insect that produced manna, either directly by excretion or by puncturing the bark of the oak or the tamarisk[16] (the association of *qaqqadânu* and Latin *cicada* is, however, onomatopoeiac rather than direct). *Aṣuṣimtu* and *ṣaṣuntu* (*aṣû*, "to exude" or "to go forth") are explained as manna or honeydew produced by the evacuation of *cicadae*.[17]

The Assyrian *šakiru* (Sumerian, "earth of the sun-god") refers to "sweet manna." There is also an etymological connection with "intoxication" and "fermentation."[18] From *šakiru* are derived the Syro-Persian *šekar*, Sanskrit

---

[12] Langdon, 1931: pp. 97–98.
[13] Fahir İz and Hony, 1952; Steingass, 1957.
[14] R. C. Thompson, 1949: p. 274; also *ibid.*, 1937: p. 228 (*ḳudratu, qudratu*). Cf. Ainsworth (1837), 1868: p. 501 (Around Sulaymānīyah, Kurdistan, "two kinds of manna (*küdrat halvassi*, "divine sweetmeat") are collected – one from the dwarf oak and another from the rocks, the latter being pure and white. When a night is unusually cool in June, the Kurds say it rains manna, as most is then found.").
[15] R. C. Thompson, 1949: pp. 268, 271 ("drug of meal" – meal-like appearance or manna added to meal?); cf. *ibid*: p. 273 (Aramaic *suphlê*, "scrapings").
[16] R. C. Thompson, 1937: pp. 229–230 (*ḳaḳḳadânu*); 1949: pp. 268, 273, 276–279.
[17] Cf. Küchler, 1904: pp. 48–49.
[18] R. C. Thompson, 1937: p. 229; 1949: p. 274.

sakkara, Latin *saccharon*, and thus "sugar."[19] Several kinds of manna were used in Persian sweetmeats long before the introduction of cane sugar.

C. EGYPT

"Manna" (*mnn, mannu, mennu*) is represented in Egyptian hieroglyphs.[20] It is found among various offerings in the Ptolemaic temple of Horus (237–257 B. C.) at Edfu or Idfu, where manna (Fig. 2) is also described as "white" or "bright" (*mennu-t ḥet'*) and compared to grains of *ānt* – apparently an Arabian or East African resin.[21] *Mennu* appears to be a loan word, from the archaic Hebrew *mân*.[22] The confusion between medicinal and comestible mannas on the one hand, and incense "manna" (*manna thuris*) on the other, may go back to Ptolemaic times. However the use of manna in medicine antedates the Ptolemaic period. The important Berlin Papyrus of ca. 1350 B.C. includes two prescriptions in which manna is an ingredient.[23]

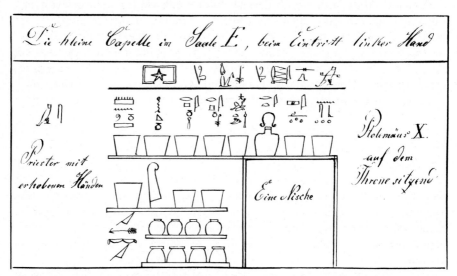

Fig. 2. Hieroglyph for "manna" (fourth row from bottom, far left), Ptolemaic temple of Horus, Edfu (Dümichen, 1867: 1: Taf. LXXXIII).

---

[19] See Levey, 1966: p. 284 (*sukkar*). Elsewhere Levey (1973: p. 7) remarks that in ancient Mesopotamia "vegetable honey" was employed as a disinfectant and as a styptic.

[20] Vocabularies of Brugsch (1867–1880: 1, 2: p. 642) and Pierret (1875: pp. 212–213), *mennu ḥet'*, "manne blanche, quelquefois tamaris." Cf. Ebert, 1908: p. 428 (*mnn*); Erman and Grapow, 1906–1931: 2: p. 71 (*mnw*).

[21] Brugsch and Dümichen, 1862–1865: 4: Taf. 88, col. 27; Dümichen, 1867: 1: Taf. LXXXIII. The inscriptions relating to manna were first discussed by Ebers, 1872: pp. 226–227. For *ānt*, see Naville, 1894: pp. 21, 24, 25 (*anti* [incense] from Punt); Ebers, 1875: 2: p. 9 (*āntí*).

[22] For the Egyptian connection, see Bochart, 1692: 3: p. 59.

[23] Deines *et al.*, 1954–1973: 4: pp. 106 (manna and sweet beer), 164 ("frische Dickmilch vom Rind; Honig; werde gegessen vom Manne an vier Tagen"). see also Ebbell (*Papyrus Ebers*, ca. 1550 B.C.), 1937: pp. 31 (VII) ff. (*wᶜḥ*), and Keimer, 1943: pp. 279–280 on the meaning of *wᶜ.ḥ.*.

D. THE MEDITERRANEAN WORLD

Greek and Latin authors used the words μάννα and *manna* ("a vegetable juice hardened into grains") to describe the aromatic and medicinal gum-resin olibanum or frankincense (Latin *tus* or *thus*), obtained from *Boswellia carterii* Birdw. and *B. frereana* Birdw. Pliny (A. D. 23–79) observed that "the fragments knocked off by striking the [frankincense] tree we call manna."[24] This was also known as *libanou manna*[25] (Hebrew *lebonah*, Greek *libanos* or *libanotos*, Latin *libanus*, Arabic *lubān* – [white] incense, frankincense).[26] Dhofâr (southern Arabia) and Somaliland were the chief areas of supply.[27] "*Thus* grows in Arabia, which is called Thurifera," according to Dioscorides (ca. A.D. 78).[28] Ptolemy (ca. A.D. 150) placed *Libanotophoros sive Thurifera Regio* in the hinterland of Dhofâr.[29] The Greek physicians Oribasius of Pergamum (ca. A.D. 325–400)[30] and Paulus of Aegina (7th century),[31] and Johannes Platearius (12th century),[32] a member of the celebrated medical school of Salerno, also mention "manna of incense." The same substance was described as "manna thuris" or "manna thurisera" in the many herbals and works on *materia medica* that appeared between the 16th and 18th centuries.[33] The 1567 edition of *El Ricettario Fiorentino* (Colegio de' Medici) refers to the "manna of the Greeks" as "la parte minuta dell' incenso.[34]

Various saccharine substances – among the "mannas" of later periods – were known to the Classical authors as "honey oil" (*elaeomeli*) or "aerial honey" (*mel ex aëre, melleus humor*). "Honey is what falls from the air,"[35]

---

[24] Pliny, 1961–1968: 4: p. 44 ("micas concussu elisas mannam vocamus"); 8: p. 258 ("turis manna una"). For ancient and medieval references to frankincense, see Birdwood, 1871: pp. 111–148.

[25] Matthiolus [Dioscorides], 1558: pp. 71–73; Dioscorides [*Herbal*, A.D. 512], 1934: p. 46; 1952–1959: 3: pp. 49–51, 58, 230, 439. Cf. Colebrook, 1807: pp. 377–382.

[26] The apparent, but erroneous, connection with the Lebanon has led several authors astray. In Chinese, frankincense = *ju-hsiang*, "milk perfume" or "milky incense" (Bretschneider, 1871: p. 19 n. 1; Wheatley, 1959: p. 47). Cf. Egyptian *mennu-t ḥet' (supra* p. 6).

[27] Beek, 1958a: pp. 141–144; 1958b: pp. 139–142; Miller, 1969: pp. 102–104, 107.

[28] Dioscorides, 1934: p. 45. Cf. Strabo (ca. 63 B.C.–A.D. 24), 1960–1969: 7: p. 333; Schoff (ed. *Periplus*), 1912: pp. 33, 120. Hasselquist (1766: p. 250) misidentifies the source of frankincense and appears to associate *thus* with "Thur or Thor [Tor], a harbour in the northern bay of the Red Sea [the Gulf of Suez]."

[29] Ptolemaeus, 1540: p. 114, and Tab. Asiae VI.

[30] Oribasius, 1851–1876: 3: pp. 604–605 ("poudre d'encense," "nous faisons uniquement usage de la manne brûlée ...").

[31] Paulus Aegineta, 1844–1847: 1: p. 448; 1914: p. 683 (*manna libani*).

[32] Platearius, 1913: pp. 147–148, 211, 225–226.

[33] Fuchsius, 1535: pp. 7–9; Sylvius, 1548: p. 87 (*mannam thuris, micas thuris*); Palea and Bartholomaeus, 1550: pp. 133, 251; Belon, 1553: p. 9b; Actuarius, 1556: p. 483; Lobel, 1576 [1570–1571]: p. 23 (*mannae thuris libani*); Dalechamps, 1586–1587: 2: pp. 1753, 1755; Wecker, 1617: p. 382; J. Bauhin, 1650–1651: 1: p. 200; Deusingius, 1659: pp. 2–3; Johnstone, 1662: p. 349; Salmasius [1588–1653], 1689: 2: p. 246; Bochart, 1692: 3: p. 875; Pomet, 1694: 1: pp. 269–270; Loeches, 1728: p. 112; Geoffroy, 1741: 2: p. 582; Savary des Bruslons, 1742: 2: p. 1185.

[34] *Ricettario Fiorentino*, 1567: p. 44; see also *ibid.*, 1548: p. 21; 1597: p. 47.

[35] Aristotle, 1965–1970: 2: pp. 191–192. Comments by Johnstone, 1662: p. 333; Bochart, 1692: 3: p. 878. Cf. Pliny, 1961–1968: 3: p. 451 ("honey comes out of the air .... at early dawn the leaves of trees are found bedewed with honey.")

in other words a kind of dew (*ros*) that settled on plants and trees and was also gathered by bees. Nowhere do we find "manna" understood as an exudation or as the excretion of an insect. In fact, the ancient writers display little direct knowledge of the substance and generally only report what was said to be found or available in the Orient, from Asia Minor to India. Small quantities were probably imported and used in medicinal compounds, but the extant works of the great physicians (Hippocrates, Dioscorides, Galenus) provide only minimal evidence of this. Theophrastus (ca. 372–288 B.C.) and Virgil (70–19 B.C.) distinguished between the common ash and the manna ash (μελία, *ornus*),[36] but the manna of *Fraxinus ornus* L., later the chief official manna of Europe, does not appear to have been observed.[37]

(a) The early Greeks and the Persians

Herodotus (5th century B.C.) mentions "the town of Callatebus [in Lydia, eastern Asia Minor] where craftsmen made honey out of wheat and tamarisks (*méli ek myríkes*)."[38] Opinions differ over what is meant.[39] He may refer to a sweetmeat of wheat flour and manna, best known from Persia (*gaz-angubīn*).[40] Elsewhere in Herodotus we read of "the Gyzantes [of western Libya] where much honey is made by bees, and much more yet (so it is said) by craftsmen."[41] This again could imply either confectionary prepared from a fruit syrup or some kind of refined manna. In a gargantuan bill of fare, engraved on a column in the palace of the Persian king Cyrus (550–529 B.C.) and said to have been read by Alexander, there appeared, according to Polyaenus (A.D. 163), "of fluid honey a hundred square *palathae*, containing the weight of ten *minae* [each],[42] perhaps a further reference to tamarisk manna.

Hippocrates (ca. 460–ca. 377 B.C.) includes *mel cedrinum* in a prescription for ulcers,[43] which appears to indicate knowledge of *manna cedrina* or *ros libani*, collected from *Cedrus libani* Barrel.[44] "A honey-like juice,"

---

[36] Theophrastus, 1961–1968: 1: p. 233; Virgil, 1967–1969: 1: pp. 46, 54, 120, 124, 336, 428, 518; 2: pp. 222, 242, 244. Dioscorides (1829–1830: 1: p. 108; 2: p. 389) also refers to μελία. Cf. Fraas, 1845: p. 156; Lenz, 1859: pp. 509–510; Abbe, 1965: pp. 161, 163.
[37] There is apparently no etymological connection with μέλι, "honey" (see Carnoy, 1959: p. 175).
[38] Herodotus, 1966–1969: 3: 345. Cf. Aristotle [attrib.] *De Mirabilibus* ..., 1963: p. 245 ("They say that in Lydia much honey is collected from trees, and that the inhabitants make small balls out of it without wax"; and again "In certain parts of Cappadocia they say that honey is made without wax and that it is of the consistency of oil.")
[39] Kolb, 1892: p. 1; Hooper, 1909: p. 31; Laufer, 1919: p. 348; Haupt, 1922: p. 236.
[40] Ouseley, 1819–1823: 1: p. 381 n. 69.
[41] Herodotus, 1966–1969: 2: pp. 396–397.
[42] Polyaenus, 1793: p. 152.
[43] Hippocrates, 1825–1827: 3: p. 316.
[44] Noted by Johnstone, 1662: p. 334; G. Bauhin, 1671: p. 497; Geoffroy, 1741: 2: p. 584; I. E. Fabri, 1776: p. 104. Belon (1555: p. 129; followed by Fothergill 1744: p. 91) confuses this with the (tamarisk) manna of Sinai.

found particularly on the oak, was known to Theophrastus,[45] who may also have been aware of the manna produced by the thorny shrubs *Alhagi maurorum* Desv. and *A. camelorum* Fisch. "From the country called Aria [eastern Khorāsān and western Afghanistan]," he observed, "there is a thorn on which is found a gum resembling myrrh in appearance and smell, and this drops when the sun shines on it."[46]

(b) The Graeco-Roman world

There are several later references to oak manna, known after ca. 1600 from Kurdistan and Luristan. Firstly, among the Greeks, we have Diodorus of Sicily (1st century B.C.) who wrote of Hyrcania, to the south and south east of the Caspian Sea: "There is a tree like an oak in appearance, from the leaves of which honey drops; this some collect and take their pleasure from it abundantly."[47] Strabo (ca. 63 B.C.–A.D. 24) mentions Hyrcania but not explicitly the oak.[48] Athenaeus of Naucratis in Egypt (late 2nd or early 3rd century A.D.) quotes from Amyntas's *Itinerary in Asia* (now lost):

"They gather [the oak manna], leaves and all, and press it in a mass, moulding it like a Syrian cake or fruit, or in some cases making balls of it. And when they are about to eat it, they break off portions from the mass into wooden cups which they call *tabaitai*, and after soaking it and straining it off they drink [the syrup]. And it is as if one soaked honey in wine and drank it, but very much pleasanter than that."[49]

Virgil refers to *roscida mella*, "distilled from the oak,"[50] and there is a similar line in Ovid's *Metamorphoses*.[51] Hyrcania is again noticed by Curtius Rufus (1st century A.D.). In his *History of Alexander the Great* we read of "a tree resembling the oak, whose leaves during the night are thickly suffused with honey; but it can be collected only before sunrise, for a slight

---

[45] Theophrastus, 1961–1968: 1: pp. 201–203. Remarked by G. Bauhin, 1671: p. 495. Cf. the discussion in I. E. Fabri, 1776: pp. 98–102. Deerr (1949–1950: 2: p. 520) states that "oak manna from *Quercus vallonia* (*Q. vallonea* Kotschy = *Q. cerris* L.) and *Q. persica*, caused by puncture of an insect, and still collected in Khurdistan, was known to the Greeks and Romans". For the former, he cites only Hesiod (ca. 800 B.C.) who, however, only associates the oak with bees' [honey](1967: p. 21, lines 233–234). Cf. Pliny, 1961–1968: 4: pp. 408–409 ("Valonias also produce mistletoe, and honey as well according to Hesiod, and it is an accepted fact that honey dew falling from the sky, as we said, deposits itself on the leaves of no other tree in preference to the Valonia oak.") For oak manna, see *infra* pp. 54–59.

[46] Theophrastus, 1961–1968: 1: p. 321. The editor, A. Hort, identifies the thorn as *Commiphora [Balsamodendrum] mukul* Engl., the myrrh tree (*bdellium*). One of the species of *Astragalus* (source of both gum and manna) might also be suggested.

[47] Diodorus, 1958–1967: 10: p. 91. Rawlinson (1839: p. 104) associates Diodorus's observation with the excretion or honeydew of an insect on the leaves of the oak.

[48] Strabo, 1960–1969: 1: p. 273; 5: p. 251; also, apparently, the provinces of Matiana (Media), Sacasene and Araxene (Armenia).

[49] Athenaeus, 1957–1967: 5: p. 237. Cf. C. Müller (ed.), 1846: p. 135 (*De Asiae Mansionibus*).

[50] Virgil, 1967–1969: 1: p. 30 ("et durae quercus sudabunt roscida mella.")

[51] Ovid (43 B.C.–ca. A.D. 17), 1966–1968: 1: p. 11 ("flavaque de viridi stillabant ilice mella.")

tepidity causes it to evaporate."[52] All the accounts of Hyrcania seem to be based on a report by Onescritus who accompanied Alexander to Asia.[53]

"India" is mentioned by several authors. Seneca (ca. 4 B.C.–A.D. 65) appears to confuse reports of manna and of cane sugar or possibly "bamboo sugar" (*tabaxir*):[54]

"For some authorities believe that bees do not possess the art of making honey, but only of gathering it; and they say that in India honey has been found on the leaves of certain reeds [*mel in arundinum foliis*], produced by a dew [*ros*] peculiar to that climate, or by the juice of the reed itself, which has an unusual sweetness and richness. And in our own grasses too, they say, the same quality exists, although less clear and less evident..."[55]

The approximately contemporaneous geographer Pomponius Mela, writing of India, used the phrase "ut in eo mella frontibus defluant,"[56]; and Claudius Aelianus (3rd century A.D.) reported:

"During the springtime in India it rains liquid honey, and especially in the country of the Prasii [Prasiaea, the upper valley of the Ganges]; and it falls on the grass and on the leaves of reeds in the marshes, providing wonderful pasturage for cattle and sheep. And the animals feast off the food with the greatest delight, for the shepherds make a point of leading them to spots where the honeyed dew falls more plentifully and settles. And they in return feast their herdsmen, for the milk which the latter draw is of the utmost sweetness and they have no need to mix honey with it as the Greeks do."[57]

Galenus (ca. A.D. 129–200) distinguished between bees' honey and "roscidum mel, aut aërium,"[58] possibly tamarisk manna. Diodorus too may allude to the tamarisk in the observation that the "Arabs who are called Nabataeans [have] plenty of so-called wild honey from trees, which they drink with water."[59] Certainly Josephus, commenting upon the celebrated manna of Sinai, available "to this very day," must refer to the tamarisk.[60]

There are other reports that appear to imply knowledge of the manna of tamarisk or of *Alhagi* sp., or possibly of both. Oribasius mentions "melea armenia" and again "melae persicae folia ejus et cimas amaram possident qualitatem ...."[61] Dioscorides' *oleo mastichino* was interpreted by the

[52] Curtius Rufus, 1821: 2: p. 89.
[53] See Pliny (1961–1968: 4: p. 25) who, however, refers not to the oak, but to "trees resembling the fig, named *occhus*." Cf. Herbert (1627–1629), 1928: p. 169. *Occhus* has been variously and very tentatively identified as *Alhagi camelorum* Fisch. and as *Calotropis procera* R. Br. (Royle, 1837: p. 107; Langkavel, 1866: p. 9; Dymock, 1890–1893: 1: p. 418), both manna-producing species. The matter remains in doubt. The river on which stands Herāt was known as the Ochus (now the Hari Rūd).
[54] Cf. Pliny, 1961–1968: 4: p. 23.
[55] Seneca, 1962–1967: 2: p. 279.
[56] Mela, 1967: p. 70.
[57] Aelianus, 1958–1969: 3: pp. 217–218.
[58] Galenus, 1530: p. 106; 1821–1823: 10: pp. 80–81.
[59] Diodorus, 1958–1967: 10: p. 91.
[60] Josephus, 1967–1969: 1 [IV]: pp. 331–335.
[61] Oribasius, 1851–1876: 6: p. 489; 1940: p. 103.

sixteenth-century Spanish commentator Andrés de Laguna as *tereniabin* (Persian *tar-angubīn*), "una especie de manna liquide" (probably of *Alhagi* sp.).[62] On the other hand, the Syrian *elaeomeli* of both Dioscorides and Pliny seems to have been the product of a species of palm.[63] Pliny also describes the "juice" of the willow (*Salix* sp.),[64] later known as manna.

---

[62] Dioscorides, 1952–1959: 3: pp. 37, 176–177; 1829–1830: 3: p. 453.
[63] Dioscorides, 1952–1959: 3: p. 35; Pliny, 1961–1968: 4: pp. 310–311.
[64] Pliny, 1961–1968: 7: pp. 45–47. The best known manniferous species is *Salix fragilis* L. (*infra* pp. 59–63).

3 MANNAS OF WESTERN AND CENTRAL ASIA AND OF NORTH AFRICA

A. ALHAGI spp.

(a) Species of Alhagi

Scientific nomenclature and distribution

Manniferous members of the genus *Alhagi* Tourn. ex Adans. are usually referred to three species:

Firstly, *A. maurorum* Medic. (1787) Desv. (1813), the *Hedysarum alhagi* of Linnaeus (1753)(Fig. 3). Synonyms (or varieties) include *A. mannifera* Desv., *A. karduchorum* Boiss. et Haussk., *A. assyriacum* Nab., and *A. napaulensium* D.C.
Secondly, *A. camelorum* Fisch. (1812), including *A. persarum* Boiss. et Buhse, *A. khirghisorum* Schrenk., *A. turcorum* Boiss., and *A. [Hedysarum] pseudo-alhagi* Desv. (M.B.).
Thirdly, *A. graecorum* Boiss., including *A. mannifera* Jaub. et Spach.

A. maurorum is, however, a *nomen ambiguum*. E. Guest and G. C. Townsend (1974) place under *A. graecorum* Boiss. – *A. mannifera* Desv. (1813: *nomen nudum*), *A. maurorum* (non Medic.) D.C., *A. maurorum* var. *karduchorum* Boiss., *A. maurorum* var. *assyriacum* Nab., *A mannifera* var. *karduchorum* (Boiss.) Keller and Shaparenko (1933); and under *A. camelorum* Fisch. – *H. alhagi* L., *A. maurorum* Medic., *H. pseudo-alhagi* M.B. (1808), *A. pseudo-alhagi* Desv. (1813: *nomen nudum*), *A. turcorum* Boiss., *A. persarum* Boiss. et Buhse, and *A. camelorum* var. *turcorum* Boiss.[1]

Recognition and use of *Alhagi* spp. can be traced to remote historical times (*infra* p. 17). *Agul* or *algul* was rediscovered by Leonhard Rauwolf (1573–1576) in the Levant.[2] Joseph Pitton de Tournefort (1717) first described the plant from the Aegean island of Syra (Síros) and "made a

---

[1] Guest and Townsend, 1966–1974: 3: pp. 499, 502.
[2] Rauwolf [first published 1581], 1693: pp. 84, 152. Tournefort ([1717] 1741: 2: p. 4) gives 1537, which appears to have misled Fothergill (1746: p. 88) and Guest and Townsend (1966–1974: 3: p. 496). Chevalier (1933: p. 278) has "Rauwolf en Egypte en 1582."

particular genus of it, under the name of *Alhagi* [*maurorum* Rauwolf]."[3]
Linnaeus (1753) preferred *Hedysarum* [*alhagi*]; in the *Critica Botanica*
(1737), *Alhagi* Tourn. is listed among the "barbarous names."[4] *Alhagi* was
restored to generic status by Michel Adanson (1763).[5] The species or
varieties on which manna is found are not often identified with certainty,
and very rarely so in the early literature. *A. maurorum* (*camelorum*,
graecorum) is a deep-rooted xerophytic shrub, up to 150 centimetres high,
and with spines one to 2.5 centimetres long. It is found in open situations on
gravels and sandy soils. The approximate distribution of *A. maurorum*, *A.
camelorum* and *A. graecorum* is shown in Map 2 (Appendix A). It comprises
a vast semi-arid region, from south east Europe and North Africa to western
and central Asia and northern and central India.

Folk nomenclature and etymology

From at least the 11th century, plants belonging to the genus *Alhagi* were
known by two Arabic names. One is *al-ḥāj*, "the pilgrim" (thus Italian
"manna dei pellegrini"). Authorities include al-Bīrūnī of Khiva,[6] Ibn al-
'Awwām (*al-hadji*),[7] Maimonides (*al-ḥāǧ*),[8] and Ibn al-Baiṭār (*hâdj*),[9] who
also give the alternative description *al-'āqūl*, "the thorn." European scholars of the 17th and 18th centuries often refer to one or both of these
names.[10] According to R. Campbell Thompson, Aramaic *'âgâ* (*A.
maurorum*) probably corresponds to Akkadian *igi*, *egu* (*ašagu*, "thorn").[11]

Al-ḥāj or al-'āqūl is chiefly used to describe *A. maurorum* (in North
Africa and South West Asia).[12] The more northerly and easterly *A.*

---

[3] Tournefort, 1741: 2: p. 4. Cf. the "thorny shrub" on neighbouring Tinos noted by Wheler,
1682: p. 52; Zallones (*Voyage à Tine*, 1809: p. 50) was unable to locate the species. Tournefort
(*op. cit.* p. 5) found no evidence that the plant yielded manna. I am unable to identify the
manna graecorum of Vincentius (ca. 1250), 1494: p. 48, and of Fuchsius, 1535: pp. 7–9. Felix
Fabri ([1483] 1892–1893: 2: p. 544), writing of the (tamarisk) manna of Sinai, added that "it is
likewise found in some parts of Greece," but this too is unconfirmed. See also *infra* p. 86 n. 3
(*Cistus* sp.).
[4] Linnaeus, 1938: p. 40.
[5] Adanson, 1763: 1: p. 328.
[6] Al-Bīrūnī (973–1048), 1973: pp. 309–310.
[7] Ibn al-'Awwām (12th century), 1864–1867: 1: p. 458.
[8] Maimonides (1135–1204), 1940: p. 82.
[9] Ibn al-Baiṭār (1197–1248), 1877–1883: 1, 1: pp. 308–309, 392–393, quoting Is-Ḥāq ibn
'Imrān. Cf. 'Abd ar-Razzāq of Algiers (18th century), 1874: pp. 342–343; Sickenberger, 1890:
p. 10; Brockelman (*Lexicon Syriacum*), 1928: p. 219 (*ḥāgtā*); Renaud and Colin (*Tuḥfat
al-aḥbāb*, 17th century), 1934: p. 87 (*ḥaǧ*).
[10] Bochart (1599–1667), 1692: p. 872 (*alchag, akul*); Cotovicus, 1619: p. 412 (*agul*);
Deusingius, 1659: p. 19 (*al-hhâgi*); Johnstone, 1662: p. 337 (*agul, alhagi*); Salmasius, 1689: 2:
p. 249 (*alha[a]gi*); G. Bauhin, 1671: p. 497 (*agul, alhagi*); Michaélis, 1774: p. 37 (*algul*);
Forskål, 1775a: p. 136 (*aghûl*); Büsching, 1775: p. 42 (*agul*). Commentaries in Délile, 1812: p.
10 (*a'âquoul*); Dinsmore and Dalman, 1911: p. 34; Ducros, 1930: p. 128; Löw, 1967: 2: pp.
414–416.
[11] R. C. Thompson, 1949: pp. 184, 289. Sumerian *ú-gir* (? also gum arabic). Bedevian (1936:
no. 198) gives Turkish *mann ayran aǧ, elhaci*. Callcott (1842: p. 267) interpreted references
to "nettles" in Job 30: 7 and Zephaniah 2: 9 as species of *Alhagi*.
[12] Barth (Aïr, Sahara, 1850), 1972 (botanical glossary: *aghul*); Duveyrier (Sahara) 1864: p.
163 (*agoûl*); Musil (Arabia, 1908–1915), 1927: p. 584 (*aḳul*); Muschler (Egypt), 1912: 1: pp.
536–537 (*aqūl*); Schweinfurth (Egypt), 1912: p. 4 (*aqul*); Anon. (*Plants of Iraq*), 1929: p. 6
(*'aqûl*); Post (Palestine), 1932–1933: 1: p. 415 (*haj*); Nicolaisen (Sahara), 1963: p. 177 (*laǧul*).

Fig. 3. *Hedysarum alhagi* L. (Sibthorp, 1806–1840:8: tab. 720) = *Alhagi maurorum* Medic. (1787) Desv. (1813).

camelorum is better known by the Persian name *ushtur khār, shutur khār* or *khār-i-ushtur* (*khār,* "thorn"); thus "camel thorn," *shauk al-jamal* (Arabic),[13] *huchtirālūk* (Kurdish),[14] and the *spina camelina* of 18th-century European authors. There are early references by al-Iṣṭakhrī (ca. 950)[15] and Nassir (1035–1042).[16] Alternatively we find *khār-i-buz(i),* "goats' thorn."[17]

In 1813, Edward Frederick, writing of Luristan (western Persia), reported manna (*gez*) on a plant called *gavan*.[18] F. R. Maunsell (1896) found the *gavvan,* "a low prickly shrub," in eastern Anatolia and Armenia.[19] This could have been *Alhagi* sp.[20]

In central Asia, *A. camelorum* is known as *dava oti* (Turkī), "camel weed,"[21] *yántaq* (Turkī, of Turfan in Sinkiang),[22] and *djantak* (Mongol).[23] G. de Meyendorff (1820) gives *tikan* for the region around Carchi (Karchi, south east of Bukhāra).[24]

Alhagi maurorum is the Sanskrit *yavāsa*[25] and *durálabhá*.[26] The names

---

[13] Watt, 1889–1893: 1: p. 165; Kirtikar and Basu, 1918: 1: pp. 421–422 (*shoukul-jaimal*).

[14] Guest and Townsend, 1966–1974: 3: pp. 496, 503.

[15] Al-Iṣṭakhrī, 1845: p. 117 (*uschtergas,* Khorāsān). Cf. Tschihatchcheff, 1853–1869: 2: p. 356 n.; Brandis, 1874: p. 144 (*kas* in Sindian).

[16] Nassir, 1881: p. 270 (*choutour ghaz,* Merv). See also Kampfer, 1712: p. 725 (*chari sjutur, aru sjirin*); I. E. Fabri, 1776: pp. 121–122 (*chari schutar*); Ainsley, 1826: 1: p. 210 (*khar-shooter,* Samarqand/Khorāsān); Burnes, 1834: 2: p. 167 (*khari-shootur,* Bukhāra); Masson, 1843: p. 455 (*kâr-shútúr,* Persia); Lehmann (1841–1842), 1852: pp. 19, 248–249 (*tschutur-char,* Bukhāra); Griffith, 1847: p. 358 (*kan shootur,* Afghanistan); Vambéry, 1868: p. 241 (*khari shutur,* Turkestan); Schlimmer, 1874: p. 357 (*khare cho-tor,* Persia); Aitchison, 1886–1887: p. 59 (*shuthar-khar,* northern Baluchistān/Khorāsān); Burkill, 1909: p. 26 (*shuthar-khar,* Baluchistān); Laufer, 1919: p. 345 (*xar-i-šutur,* Persia); Hooper and Field, 1937: p. 81 (*kār shūtur,* Persia). *A. maurorum* or *H. alhagi* is similarly named by Pottinger, 1816: p. 102 (*kharé shootoor*); Roxburgh, 1820–1832: 3: p. 344 (*shooturk*); Royle, 1839: 1: p. 194 (*ooshtur khar*).

[17] Aitchison, 1886–1887: p. 467; 1888–1894: p. 59; Dymock, 1890–1893: 1: p. 418; Dragendorff, 1898: p. 326; Burkill, 1909: p. 26 (Baluchistān; also *shinz* [cf. Pottinger, 1816: p. 102] and *ghaz*).

[18] Frederick, 1819: pp. 253, 256. *Gez* he took to be the excretion of an insect.

[19] Maunsell, 1896: p. 233 (used for firewood; without reference to manna). Frederick 1819: p. 253) also mentions the vicinity of Moosh (Muṣ), eastern Anatolia.

[20] Also identified as *Tamarix gallica* (Hooper, 1900: p. 32); but Frederick stated that "tamarisk bears no resemblance to the *gavan.*" Cf. Tabeeb, 1819: p. 268; Polak, 1865: 2: p. 287 ("Gevenn" = *Astragalus* sp.).

[21] Gilliat-Smith and Turril, 1930: 8: p. 376.

[22] Le Coq, 1911: p. 99 (cf. Laufer, 1919: p. 345 n. 4); thus *yántaq-šākārī,* "alhagi-sugar" (manna).

[23] Bretschneider, 1898: p. 987 ("common in the desert around Ha mi and Sha chou, on the Lob nor, Tarim, in east Turkestan.").

[24] Meyendorff, 1826: p. 206.

[25] Roxburgh, 1820–1832: 3: p. 344; Watt, 1889–1893: 1: p. 165 (*girikarnika-yavása*); Monier-Williams (*Sanskrit-English Dictionary*), 1899 (*yāsa, yavāsa*); Dutt, 1900: p. 145; Kirtikar and Basu, 1918: 1: pp. 421–422 (*yavása, girikarnika*); Bhishagratna (*The Sushruta Samhitā*) 1907–1918: 4: p. 81; Meyerhof, 1947: p. 34; Gopal, 1964: p. 67; Singh and Chunekar (*Bṛhattrayī*), 1972: pp. 328–329.

[26] Dymock, 1885: 2: pp. 178–181; Watt, 1889–1893: 1: p. 165; Dey, 1896: pp. 16–17 (*durlavá*); Dutt, 1900: p. 145; Bhishagratna (*The Shushruta Samhitā*), 1907–1918: 1: p. 456; 4: p. 58; Kirtikar and Basu, 1918: 1: pp. 421–422 (*durlabha*). Dragendorff (1898: p. 326) gives *durálabha* (Sanskrit) under *A. camelorum,* and similarly Hindi *jawása.*

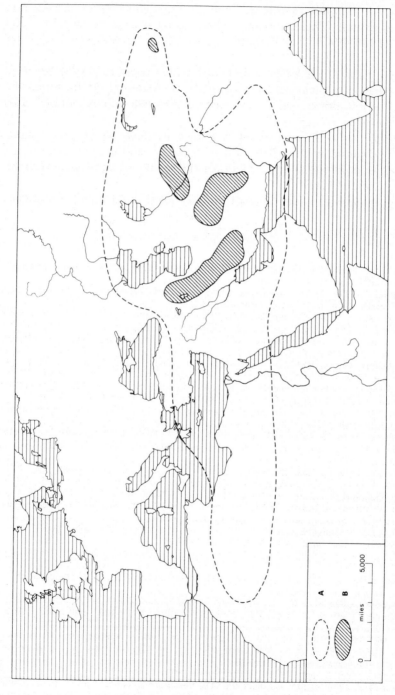

Map 2. A: Combined distribution of *Alhagi maurorum* Desv., *A. camelorum* Fisch., and *A. graecorum* Boiss. B: Collection of alhagi manna.

survive as *jawāsa* (Hindi)[27] and *dulal-labhá* (Bengali).[28] According to H. W. Bellew, *H. alhagi* is called *zôz* in Pushtu (north east of Peshawar);[29] in parts of Sindh the same plant is known as *kas*.[30]

(b) Uses of Alhagi spp.

Alhagi spp. have been put to a variety of uses. They are valued as ornamentals on account of the handsome crimson flowers. In western and central Asia the young shoots are widely consumed by grazing animals. "After all other shrubs and plants have dried up owing to the autumnal hot winds, [*A. camelorum*] still remains a vivid green, and is eagerly sought for as fodder by camels, donkeys and goats."[31] J. Schlimmer (1874) was told that in parts of Persia shepherds were required to keep their flocks of sheep and goats out of areas of manna-producing alhagi.[32]

An oil prepared from the leaves of alhagi, as well as the dried leaves, flowers, twigs, and the expressed juice were employed in folk medicine.[33] The roots of *A. maurorum* are dried and ground into flour by the Tuareg of Fezzan.[34] R. Mignan (1839) reported the curious custom in northern Persia of planting water melon seeds in the divided stems of the plant. "The seed becomes a parasite, and the nutritional matter, which the brittle, succulent roots of the melon are ill-adapted to collect, is abundantly supplied by the deeper searching and tougher fibres of the root of this thorn. An abundance of good water melons is thus periodically forced from saline soils incapable of other culture."[35]

In India, screens (*tatties*) are made from the branches.[36] While Sanskrit sources (early centuries A.D.) mention *A. maurorum*, there is no direct reference to manna. However a medicinal substance prepared by evaporating a decoction of *A. maurorum* was known as *yavāsa*-sugar. L. Gopal observed that "*yavāsaśarkarā* is another variety of sugar mentioned by Suśruta and Caraka. It was most likely extracted from the *yavāsa* plant mentioned in the *Uṇādi-gaṇa-sūtra*."[37] Whether the saccharine property is related to manna (not reported from India on *Alhagi* spp.) does not appear to have been considered.

[27] Roxburgh, 1820–1832: 3: p. 344; Royle, 1839: 1: p. 194; Aitchison, 1869: p. 44; Brandis, 1874: p. 144; Dymock, 1885: 2: pp. 178–181; 1890–1893: 1: p. 418; Dufrené, 1887: p. 8; Dey, 1896: pp. 16–17; Bamber, 1916: p. 79.
[28] Watt, 1889–1893: 1: p. 165; Kirtikar and Basu, 1918: 1: pp. 421–422.
[29] Bellew, 1864: p. 238.
[30] Brandis, 1874: p. 144.
[31] Aitchison, 1891: p. 8, writing of western Afghanistan and northeastern Persia. Note the early observations by Kämpfer, 1712: p. 725 ("spina camelina propter ea nuncupata quod in desertis camelos satiet.").
[32] Schlimmer, 1874: p. 357.
[33] Hallé, 1787: pp. 673–674; Watt, 1889–1893: 1: p. 165; Dymock, 1890–1893: 1: 418–419.
[34] Duveyrier, 1864: p. 163; Trotter, 1915: p. 180; Chevalier, 1933: p. 281; Nicolaisen, 1963: p. 177 (Tasilé-n-Ajjer).
[35] Mignan, 1839: pp. 214–216.
[36] Roxburgh, 1820–1832: 3: p. 344; Irwin, 1839: p. 892; Brandis, 1874: p. 145; Watt, 1889–1893: 1: p. 166; Guest and Townsend, 1966–1974: 3: p. 496.
[37] Gopal, 1964: pp. 67–68. See also Monier-Williams (*Sanskrit–English Dictionary*), 1899 (*yavāsa-sarkarā*); Bhishagratna (*The Sushruta Saṃhitā*), 1907–1918: 1: p. 456 (*yavāsa-s'arkárá*); 4: p. 81 (*yava'saka*); Singh and Chunekar (*Bṛhattrayī*), 1972: p. 328 (*yavāsaka*).

(c) Tar-angubīn

The manna of *Alhagi* spp. is a spontaneous exudation in the form of light brown globules or "tears" on the stem and leaves of the plant. There is no convincing evidence that insects are involved.[38] Nevertheless, the substance is found only in certain areas. The reason for this is as yet undetermined, but may be expected to lie in prevailing ecological conditions, notably low rainfall, high day temperatures in summer, a large diurnal and/or seasonal range of temperature, and possibly the presence of certain soil nutrients. The manna exudes at night or around dawn (thus the early association with "dew") and is collected by simply shaking or beating the branches of alhagi over a cloth spread on the ground. The harvest lasts for three or four weeks, when the plant is in flower or the seeds are ripening, between June and August.[39] In a good season a man could collect up to 25 kilograms in a day.[40] Like several similar substances it has been associated with the manna of the Israelites.[41]

In the areas of origin and more widely in western Asia, alhagi manna was used in the preparation of syrups and sweetmeats,[42] as well as in medicinal compounds. Elsewhere, the product – containing melezitose, but not mannite[43] – was valued chiefly, and perhaps exclusively, as an item of *materia medica*. It was displayed at the International Exhibition held in London in 1862 and in St. Petersburg in 1870 (a sample from Turkestan).[44]

The Middle Ages

The manna of *Alhagi* spp. reached Europe during the Middle Ages.[45] Obtained in the seaports of the Levant and imported through Venice, it was sometimes described as "Syrian manna" or "Arabian manna" and, more accurately, as "Persian manna."[46] From the 15th century this and other Oriental mannas were gradually displaced in the *materia medica* of Europe by the product of Calabria (*infra* pp. 87–97).

---

[38] Délile, 1812: p. 10; Anon., 1828: p. 262; Burnes, 1834: 2: p. 167; Wellsted, 1838: p. 48; Griffith, 1847: p. 358.

[39] Roxburgh, 1820–1832: 3: p. 344; Burnes, 1834: 2: p. 167; Irwin, 1839–1840: p. 892; Griffith, 1847: p. 358; Vambéry, 1868: pp. 241–242; Brandis, 1874: p. 144; Alëkhine, 1889: p. 536 n.; Aitchison, 1891: p. 7; Bamber, 1916: p. 79. Insects attracted by the afflorescence might prompt the idea that they were in some way responsible for the presence of manna.

[40] Alëkhine, 1889: p. 536 n., quoting a local informant, M. Ivanoff.

[41] Büsching, 1775: p. 42; Don, 1831–1838: 2: p. 310 ("Hebrew manna"); Landerer, 1842: pp. 371–372; Haussknecht, 1870: p. 248 ("Manna Hebraïca"); Sykes, 1906: p. 433.

[42] Moghadam (1930: p. 85) doubts this, suggesting confusion with the manna of *Tamarix* spp., but the evidence is strong. At the same time, alhagi manna was probably also mixed with other, similar substances.

[43] For the chemical composition, see Ludwig, 1870: pp. 41–44; Villiers, 1877: pp. 35–38; Flückiger and Hanbury, 1879: p. 414; Flückiger, 1883: pp. 26–27; Markownikoff, 1885: p. 943; Alëkhine, 1889: p. 535; Dymock, 1890–1893: 1: p. 420; Ebert, 1908: pp. 469–470; Hudson and Sherwood, 1918: pp. 1459–1460; Wehmer, 1929: p. 350; Moghadam, 1930: p. 87.

[44] Hanbury, 1863: p. 109; 1876: p. 289; Akëkhine, 1889: p. 536.

[45] Heyd, 1886: 2: p. 633; Alëkhine, 1889: p. 535 ("miel de fruits").

[46] Pasi, 1521: pp. 90b, 92a, 115b, 187b, 189a, 190b, 192a, 193a (*manna soriana*). "Persian manna" is included in a prescription by Abū al-Qāsin (Abulcasis, ca. 936–1010) of Córdoba (Hamarneh and Sonnedecker, 1963: p. 74).

Alhagi manna is the *tar-angubīn* of Persia,[47] literally "green honey" or "moist honey," otherwise *mel humidum* and *miel de rosée*.[48] As *ṭarmâgbîn* it appears to be mentioned in the Syriac *Book of Medicines* (early centuries A.D., but incorporating earlier material).[49] Persian and Arab scholars refer to *tar-angubīn* from the 9th or 10th century. Perhaps the first was 'Ali ibn Sahl al-Ṭabarī.[50] From these statements we can identify the chief areas of collection – Persia, especially Khorāsān, and west-central Asia, notably around Bukhāra.

The anonymous author of the *Ḥudūd al-'Ālam*, "The Regions of the World," a Persian geography of A.D. 982, stated that Kish (Kishsh or Kesh) "produces good mules, manna (*tarangabīn*), and red salt, which are exported everywhere."[51] This may be the source of a similar statement by al-Idrīsī (1154).[52]

Al-Bīrūnī, born in Khwārizm (Khiva) A.D. 973 and "one of the very greatest scientists of Islam,"[53] has important passages on manna in his "Book of Drug Knowledge" (*Kitāb as-Ṣaydanah*). "Some people believe," he reported, "that [the manna of the Israelites] is *taranjubīn* and deposits on the camel's thorn ...."[54] Al-Bīrūnī quoted an unknown author, al-Fazārī, to the effect that this manna was excreted by an insect, which suggests confusion with *tréhala* (infra pp. 40–42).[55]

Ibn Sīna (Avicenna, born near Bukhāra, A.D. 980, died near Hamadān, 1037) associated *tar-angubīn* with Khorāsān,[56] and likewise Ibn Buṭlān (died 1052) of Baghdād,[57] Ibn Sarābī (? 12th century),[58] and the Hispano-Arab botanist and pharmacist Ibn al-Baiṭār (1197–1248).[59] Other references to *tar-angubīn* may be found in the works of al-Rāzī (ca. 900) of Raī near

---

[47] Vullers (*Lexicon Persico-Latinum*), 1855–1867: 1: p. 440; Steingass (*Persian-English Dictionary*), 1957: p. 297. Spanish *tereniabin* from the Arabic *terenjobīn*.
[48] Aitchison (1886–1887: p. 467) and Hooper (1909: p. 33) give "honey from the green [bush]." Later, Aitchison (1891: p. 206) thought, probably incorrectly, that the name had "been merely transferred" from *gaz-angubīn*, the manna of *Tamarix* spp. Both are collected in Persia, and the presence of fragments of leaves sometimes gives a green appearance.
[49] Budge (ed.), 1913: p. 481 (a prescription); see also pp. 332, 401, 719 (*ṭalla daghbîn*). The work is a translation from the original Greek. Chassinat (1921: p. 71) has published *Un Papyrus Médical Copte* in which there is a reference to manna, possibly of alhagi.
[50] Al-Ṭabarī, 1969: p. 291 no. 471.
[51] Anon. (trans. V. Minorsky), 1937: p. 113. Kish is now represented by Kitāb and Shahr-i-Sabz, approximately 60 kilometres south of Samarqand.
[52] Al-Idrīsī, 1836–1840: 2: p. 200 (Kech: "les montagnes [environnantes] produisent en abondance du *terendjebīn* [sorte de manne].")
[53] Sarton, 1927–1948: 1: pp. 707–709. Al-Bīrūnī died in 1050 at Ghazna (Ghazni) in Sīstan, Afghanistan. He wrote in Arabic.
[54] Al-Bīrūnī, 1973: 1: p. 310.
[55] Meyerhof, 1947: p. 33. In discussing *taranjubīn*, al-Bīrūnī also quotes Abū Ḥanīfa al-Dīnawarī, Arab philologist and naturalist of the 9th century.
[56] Ibn Sīna, 1608: 2: p. 404.
[57] Ibn Buṭlān, 1531: p. 24.
[58] Ibn Sarābī, 1531: p. 36. See also Guignes, 1905: 6: pp. 58–59 (no. 360), 84 (no. 497). I have followed Sarton (1927–1948: 2, 1: pp. 229, 608) in distinguishing between Ibn Sarābī (Serapion Junior, or the Younger, first half of the 12th century) and Yahyā [Yūhannâ] ibn Sarāfyūn (Seraphion the Elder, second half of the 9th century). It has been concluded that they were one and the same person.
[59] Ibn al-Baiṭār, 1877–1883: 1: pp. 308–309 (no. 408), quoting Is-Haq ibn 'Imrān.

Tehrān,[60] Muwaffiq ibn 'Alī (ca. 970) of Herāt,[61] Averroës (Ibn Rushd', 1126–1198) of Córdoba,[62] Maimonides (1135–1204) of Córdoba and later Cairo,[63] and al-Samarqandī (died 1222 in Herāt).[64] Although these scholars borrowed extensively from one another it is apparent that the manna of *Alhagi* spp. was available and in pharmaceutical use throughout the Arab world. Europeans of the later Middle Ages who mention manna and may refer to *tar-angubīn*, include the travellers Fr. Jordanus (between Persia and "India the Less"),[65] Fr. Odoric of Pordenone (at Huz on the western margins of Persia),[66] Ruy González de Clavijo ("India"),[67] and Joos van Ghistele (Persia and Arabia).[68]

The Chinese obtained the manna of *Alhagi camelorum* from central Asia, either directly or through Persian or Arab intermediaries. It was understood to be a kind of "sweet dew," *kan-lu*.[69] The *Annals* of the Wei (386–558) and of the Sui (589–618) dynasties "ascribe to the region of Kao-č 'aṅ [Kao ch'ang – Turfan or T'u-lu-fan in eastern Sinkiang] a plant, styled *yaṅ ts'e*, the upper part of which produces honey [*mi*] of very excellent taste."[70] Bernard Laufer argued that *yaṅ ts'e* is probably a literal rendering of a lost Middle Persian or Sogdian term for "sheep thorn,"[71] that is *A. camelorum* (also "camel thorn" and "goats' thorn"). The manna of Turfan was known as *ts'e mi*, "thorn honey,"[72] It is mentioned again in 981.[73] The Persian name *tar-angubīn*, familiar to the Chinese through trade contacts with Samarqand, was transcribed as *ta-laṅ-ku-pin*.[74] The Chinese of the 12th century were also aware that another variety of *kan-lu* (? oak manna) was collected in the vicinity of Mosul (Kurdistan).[75]

---

[60] Al-Rāzī, 1766: p. 179; Berendes, 1965: 2: p. 131.

[61] Muwaffiq ibn 'Alī, 1968: pp. 173, 355.

[62] Averroës, 1531: p. 361.

[63] Maimonides, 1940: pp. 82 (no. 166), 193 (no. 386).

[64] Al-Samarqandī, 1967: p. 91.

[65] Jordanus (ca. 1300), 1863: p. 10.

[66] Odoric (1286–1331), 1891: p. 59; 1913: p. 109.

[67] González de Clavijo (1403–1406), 1859: p. 93; 1928: pp. 159, 289 (brought to the markets of Solṭānīyeh and Samarqand).

[68] Ghistele (1446–1525), 1557: pp. 194, 277–278. See also Vincentius (ca. 1250), 1493: p. 47; Mauro (*Mappamondo*, 1457–1459, notes on Arabia), 1806: p. 48 (cf. Heyd, 1886: 2: p. 632).

[69] Equivalent to Sanskrit *amrta* or *amrita*, "ambrosia," "nectar [of the gods]." See Chavannes and Pelliot, 1911–1913: p. 131; Hirth and Rockhill, 1911: p. 140 n. 1; Stuart, 1928: p. 258.

[70] Laufer, 1919: p. 343. Repeated in the *Ming Geography* (Bretschneider, 1888: 2: pp. 192–193).

[71] Laufer, 1919: p. 343. Called by the Hu (Iranians) *k'ie-p'o-lo*, according to a Chinese authority of the 8th century; ? *k'ie* = Persian *khār*, "thorn." See also F. P. Smith, 1871: p. 144; Stuart, 1928: p. 258.

[72] Laufer, 1919: p. 344 and n. 5 (A.D. 667–730, cited in the *Pen ts'ao kaṅ mu*, 1596). Cf. Read and Liu Ju-Ch'iang, 1927: no. 342 (*tz'u mi* from *A. maurorum* Medic.); Read, 1936: p. 111.

[73] Schott, 1875: pp. 47–48. See also Schafter, 1977: p. 108 (T'ang, A.D. 618–907). Joos van Ghistele (1446–1525), who travelled in western Asia and Tartary, referred (1557: p. 194) to the manna of Persia and Cathay. By the latter he presumably meant central Asia, possibly Sinkiang.

[74] Bretschneider, 1888: 2: p. 254 (*ta-lang-gu-bin*); Laufer, 1919: p. 345.

[75] Laufer, 1919: p. 344; Hirth and Rockhill, 1911: p. 140. Concerning Mosul (?), see also Rockhill, 1915: pp. 621–622; Laufer, 1919: p. 344 n. 1. Stuart (1928: p. 259, based on F. P. Smith, 1871: p. 144) claimed that *ch'eng-ju* = "tamarisk manna," but this was refuted by Laufer, 1919: p. 348.

In the 16th century, European students of *materia medica* with no direct knowledge of the East described *"tereniabin"* as "manna arabum" or "manna orientalis." The earliest notices discovered are in works by Valerius Cordus (*Dispensatorium*, 1535)[76] and Antonio Musa Brasavolus of Ferrera (*Examen Omnium Simplicium Medicamentorum*, 1537).[77] It is not altogether clear that they refer to the manna of *Alhagi* spp. The naturalist Pierre Belon du Mons, who visited the Levant, appears to have confused *tar-angubīn* ("manne liquide ... tereniabin") with tamarisk manna, collected in Sinai and marketed in Cairo.[78] Garcia da Orta (1563), domiciled in Goa, has a more circumstantial account: "They say that [*tiriamjabim* or *trumgibim*] is found among the thistles and in small pieces, somewhat of a red colour ... they are obtained by shaking the thistles with a stick and ... are larger than a coriander seed when dried.... The vulgar hold that it is a fruit, but I believe that it is a gum or resin. ... it is much used [in medicine] in Persia and Ormuz."[79] The red colour was presumably derived from the flowers of *A. camelorum*. The observations of Christovão Acosta (1574)[80] and Jan Huyghen van Linschoten (1583–1592),[81] notwithstanding their experience of the East, owe much to Orta.[82] According to Leonhard Rauwolf (*Travels*, 1573–1576), "From Persia they bring [to Aleppo] great quantity of an unknown manna in skins, by the name of *trunschibil*, which is gathered from a prickly shrub, called by the Arabians *agul* and *alhagi*, which is the reason that it is mixt with small thorns and reddish chaff."[83] Prosper Alpinus (1591) states that *tereniabin*, red in colour (*rubra*), was exported from "Armenia."[84]

Not much new information was published in Europe in the 17th century. *Tar-angubīn* (of which probably little was known directly[85]) was described as "manna liquida"[86] (Fig. 13), "manna persiana,"[87] and even "manna indica."[88] Pedro Teixeira (*Travels*, 1586–1605) observed that *"toraniabin* is

---

[76] Cordus, 1599: p. 209. See also Fuchsius, 1535: pp. 7–9 (*mannam Arabum*).

[77] Brasavolus, 1537: p. 338. Cf. Matthiolus, 1544: p. 48; 1558: p. 73; 1598: p. 94; Scaliger, 1557: p. 226 ("In Arabia etiam liquidius colligitur manna, quam prisca voce etiamnunc vocant *tereniabin*"); Lobel (1570–1571), 1576: pp. 23–25; Altomarus (Briganti), 1562: p. 10.

[78] Belon, 1555: p. 129. Similarly Alpinus, 1591: p. 127 (*terengibil*); Cotovicus, 1619: p. 412 (*transchibal*).

[79] Orta, 1913: p. 281. At about this time manna from Persia appears to have reached Goa *via* Syria (Hakluyt, 1903–1905: 6: p. 27 [A.D. 1584]). Concerning the notion that this manna was "a fruit," see Alëkhine, 1889: p. 535 ("miel de fruits").

[80] Acosta, 1578: p. 308.

[81] Linschoten, 1885: 2: p. 100. Gaspard Bauhin ([1560–1624], 1671: p. 497) appears to have regarded the *trunschibil* of Rauwolf and the *trumgibiin* of Linschoten as different kinds of "manna persiana."

[82] Also Clusius, 1605: p. 164.

[83] Rauwolf (1581), 1693: p. 84.

[84] Alpinus, 1591: p. 127.

[85] Pomet (1694: 1: p. 239) observed that *tereniabin* was "rare in France." According to Merat and Lens (1829–1834: 1: p. 165) its use in medicine was unknown in France.

[86] Johnstone, 1662: p. 334; Angelus, 1681: p. 359. Cf. Salmasius (1588–1653), 1689: p. 249 ("Duo genera mannae ab Aribibus prodita, liquidum et concretum"). The manna of *Alhagi* spp. is moist and viscous, but not "liquid" (see Moghadam, 1930: p. 85).

[87] G. Bauhin (1560–1624), 1671: p. 497.

[88] J. Bauhin (1541–1613), 1650–1651: 1: p. 199 ("manna indica *tiriamiabin* aut *trungibim*").

found in many parts of Persia. It is very like dry coriander seed and is produced on certain herbs like wild thistles."[89] John Chardin's comments (1666–ca. 1675) are ambiguous. He refers to one kind of manna from Nichapour (Neyshābūr, Khorāsān) and a "liquid manna" collected around Ispahan.[90] Samuel Bochart (1599–1667) correctly associated the product with the spiny *alchag* (alhagi) in the territories of Khorāsān, Bactria, Sogdiana and the Oxus.[91] John Ray (1693) and Pierre Pomet (1694) added Persia and "around Aleppo."[92] *Tar-angubīn* was doubtless available in the pharmacies of Aleppo (as reported earlier by Rauwolf), but it is unlikely that it was collected locally. Alexander Russell (1756), who studied the natural history of the area, found that "what [*H. alhagi*] grows in the vicinity of Aleppo is of low growth and produces no manna."[93]

The botanist Pitton de Tournefort (1717) made a significant contribution to the study of *Alhagi* spp. (*supra* p. 12). "It is chiefly about Tauris (Tabrīz)," he added, "that [the manna] is gathered, under the name of *trungibin* or *terenjabin*, ... During the great heats, you perceive small drops of honey shed on the leaves and branches of these shrubs .... They .... make 'em into reddish cakes, full of dust and leaves .... Two sorts are sold in Persia; the best is in little grains, the other is like a paste, and contains more leaves than manna."[94] Karsten Niebuhr (1761–1764) examined a sample of *tar-angubīn* at Basra; this apparently had been brought from the region of Ispahan.[95] Among those who travelled in Persia at this time, Samuel Gmelin (1770–1772) also mentions the *thereniabin* of Ispahan.[96] According to Guillaume Antoine Olivier (ca. 1793) it was found, towards the end of summer, "dans les contrées les plus chaudes de la Perse, ainsi que dans l'Arabie," but not on the alhagi of Rhodes, Cyprus, Crete and Syria.[97] Like Niebuhr, he maintained that *tar-angubīn* was used in confectionery as well as in medicine.[98]

---

[89] Teixiera, 1902: p. 203.

[90] Chardin, 1811: 3: pp. 295–296; 1927: p. 140 (in addition to the manna found on tamarisk). Cf. Pinkerton, 1811: p. 180; Sykes, 1906: p. 433.

[91] Bochart, 1692: 3: p. 872. See also Deusingius, 1659: p. 19.

[92] Ray, 1693: 2: Appendix, *Stirpium Orientalium Rariorum Catalogus*; Pomet, 1694: p. 239.

[93] Russell, 1794: p. 259.

[94] Tournefort, 1741: 2: pp. 5–6. Tournefort (*op. cit.*, p. 4) described the alhagi as "one of the beauties of the plains of Armenia, Georgia and Persia." From this, Savary des Bruslons (*Dictionnaire universel de Commerce*, 1742: 2: p. 1185) appears to have assumed that *teringabin* was collected in Georgia and Armenia generally, as well as around Tabrīz. John Stevens (1715: p. 39) refers only to "several parts of Persia." See also Kämpfer, 1712: p. 725.

[95] Niebuhr, 1773: p. 129 ("*Tarandsjubín* ou *tarandsjubíl* se recueilloit en grande quantité dans la contrée d'Isfahán sur un petit buisson epineux."); 1792: 2: p. 360 ("As Arabia-Petraea abounds in prickly shrubs, possibly this manna may be found also there; although in those desert places it cannot be very plentiful." Niebuhr considered whether *tar-angubīn* might have been the manna of the Hebrews; similarly Hallé, 1787: pp. 397–398.

[96] Gmelin, 1774: p. 288.

[97] Olivier, 1801–1807: 3: pp. 188–189. Concerning "Arabia," see also Michaélis, 1774: pp. 37, 40.

[98] European students of *materia medica* in the 18th century chiefly attempted to clarify and, where necessary, to reconcile statements made by earlier authorities. See Geoffroy, 1741: 2: pp. 585, 600–601; Fothergill, 1746: pp. 86–87; I. E. Fabri, 1776: pp. 121–122. 'Abd ar-Razzāq (1874: pp. 342–343), an 18th-century physician of Algiers, also mentions *tarandjoûbin*.

After 1800

Information from the period after 1800 is valuable in locating the chief areas of manna production (Maps 2 and 3). These were (a) west-central Asia, between the Amu Darya (Oxus) and the Syr Darya (Jaxartes), part of the ancient *Turan*; (b) western Persia (where the risk of confusion with other mannas is most apparent), and (c) eastern Persia, western Afghanistan and northern Baluchistān. No modern evidence of alhagi manna from the Turfan oasis (Sinkiang) has been found; nor is it possible to confirm that the product was ever collected in Arabia.[99] A. Villiers examined a sample of alhagi manna said to have been "collected" in Lahore (1870);[100] with this exception and that of Baluchistān, there seems to be general agreement that the product was not locally available in the sub-continent of India.[101] Claims for parts of the Sahara[102] are unsubstantiated, and the alhagi of Egypt, Palestine and the Levant is not known to produce manna.[103] However, bearing in mind the adventitious nature of the evidence generally, it is very likely that the substance occurs more widely, unnoticed and/or unexploited, than is shown in Map 2. In particular, the three main areas may be part of a single zone in the eastern half of the distribution of *Alhagi* spp.

Central Asia: Here the localities cluster around Bukhāra, Samarqand and Tashkent. M. Alëkhine (1889, based on information supplied by M. Ivanoff) noted (a) the steppes around Karchi[104] and (b) Ouratubé (unidentified); (c) the commune of Boukine in the district of Kouramine (? Kermine, now Navoi), to the south west of Angréna (Angren);[105] (d) the steppe between Djisak (Dzhizak) and Tschinaze (Chinaz); and (e) the region of Dinaou (Deynau) – Ourtchi (? Uch-Adzhi).[106] To the east of Dzhizak lay Zaamin where, at the close of the 17th century, "L'on recueille . . . . la manne la plus

---

[99] Chevalier (1933: p. 276) maintains that it was (Il n'est pas douteux . . . . que la manne d'*Alhagi* est utilisée en Arabie et même parfois exportée."), but he cites no observations. See also Délile, 1812: p. 10; Alëkhine, 1889: p. 535 ("l'Arabie et la Palestine"). Royle (1839: 1: p. 194), Baillon (1871–1892: 1: p. 108) and Laufer (1919: p. 346) state that it was not available in Arabia. In the early literature, the "manna of Arabia" is sometimes mentioned, but with little or no indication of the kind; see Mauro (1457–1459), 1806: p. 48 (called *mechina*); Ghistele (ca. 1500), 1572: p. 311; Linschoten (1583–1592), 1885: 1: p. 47.

[100] Villiers, 1877: p. 35. Honigberger (1852: pp. 305–306) observed that there were only two kinds of manna in the bazaar at Lahore (*tooroonjebún* and *shirkesht*), both imported from Persia *via* Kabūl.

[101] Watt (1889–1893: 1: p. 165) quoted contrary observations; one indicated that "small quantities" of *taranjabin* were produced in several parts of India. Hanbury (1876: p. 289), Flückiger and Hanbury (1879: p. 414) and S.G Harrison (1951: p. 412) refer to samples that had been imported.

[102] Chevalier, 1933: p. 280.

[103] Landerer (1842: pp. 371–372) maintained that *H. alhagi* yielded manna in the Lebanon only after being browsed by goats.

[104] See also Meyendorff, 1826: p. 206; Burnes, 1834: 2: p. 167. Burnes noted that *turunjubeen* was "not found westward of Bukhāra . . . . though in general plenty to the eastward, near Kurshee [Karchi] and Samarqand." Lehmann ([1841–1842], 1852: pp. 248–249) referred particularly to the area to the south of Bukhāra and Samarqand. Kish, mentioned A.D. 982 (*supra* p. 19 n. 51), lay south of Samarqand and north east of Karchi. See also Don, 1831–1838: 2: p. 310 ("Caspian manna").

[105] Navoi and Angren are, however, far apart. Both are shown in Map 3.

[106] Alëkhine, 1889: p. 536 n.

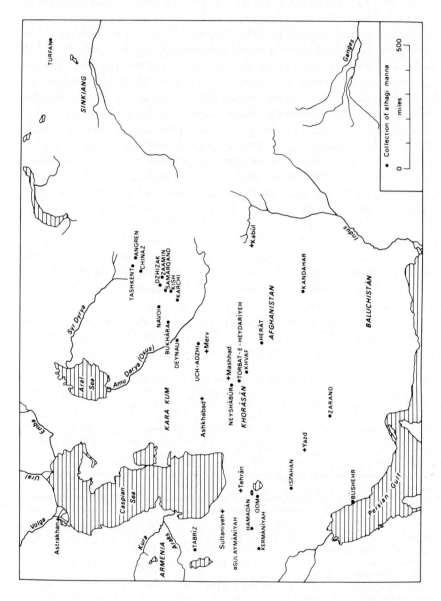

Map 3. Places named in references to the collection of alhagi manna.

24

exquise de tout l'Orient, que les Persans, et ensuite les Arabes, appelent *terengiubin alzamini.*"[107]

In the vicinity of Tashkent, according to Ivanoff, the manna was called *rousta* or *yantaq-chakar*,[108] of which there were several qualities. G. de Meyendorff, writing of Bukhāra (1820), described *rousta* as a "sirop" prepared from *terendjebin*.[109] This is the *shire* reported by A. Vambéry.[110] Tashkent had a considerable trade in alhagi manna. India obtained the product partly from Persia and partly from Turkestan.[111] Bukhāra appears to have specialized in making confectionery (*candalates*) that included refined manna. Cane sugar was expensive and manna was one of several substitutes.[112]

Persia, Afghanistan and neighbouring territories: *Tar-angubīn* from different parts of Persia was exported to India[113] and to North Africa and Egypt where it was known as *mann farssy* or *mann el hhagy*.[114] According to S. Moghadam (1930), "La manne *Alhagi* est la seule manne en larmes qui existe en Perse dans le commerce."[115] Production around Tabrīz (Āžarbāī jān) is mentioned in early sources (*supra* p. 22) and in several 19th- and 20th-century accounts.[116] It may extend into Turkish and Russian Armenia. Southward, there are reports from Kurdistan[117] (including Sulaymānīyah in north east Iraq[118]), Hamadān,[119] Kermānshāh,[120] Qom,[121] Ispahan,[122] Zarand,[123] and Būshehr (Bandar e Būshehr).[124] Vambéry (1868) observed that *terendjebin* was "used in the sugar manufactures of Yezd [Yazd] and Meshed [Mashad],"[125] the latter presumably drawing upon supplies from Khorāsān.

[107] Herbelot, 1697: p. 923.
[108] Presumably the manna called *cherker* from Little Tartary and northern Khorāsān reported by Olivier (ca. 1793), 1801–1807: 3: p. 189.
[109] Meyendorff, 1826: p. 206.
[110] Vambéry, 1868: p. 241.
[111] Kirtikar and Basu, 1918: 1: pp. 421–422.
[112] Fraser, 1826: p. 96; Burnes, 1834: 2: p. 167; Royle, 1839: 1: p. 194.
[113] Royle, 1839: 1: p. 194; Honigberger, 1852: 2: pp. 305–306; Watt, 1889–1893: 1: p. 165; Dymock, 1890–1893: 1: p. 420.
[114] Ducros, 1930: p. 128.
[115] Moghadam, 1930: p. 85.
[116] Green, 1820: 1: p. 663; O'Shaughnessy, 1842: p. 295; Collin, 1890: p. 103; Moghadam, 1930: p. 81.
[117] Watt, 1889–1893: 1: p. 165; Dymock, 1890–1893: 1: p. 419, quoting Mir Muḥammad Husain.
[118] Mignan, 1839: 1: p. 289 (traded to Baghdād and Basra where it was used in confectionery).
[119] Dupré (1807–1809), 1819: 2: p. 375; Watt, 1889–1893: 1: p. 165; Dymock, 1890–1893: 1: p. 419.
[120] Dupré (1807–1809), 1819: 2: p. 375.
[121] Haussknecht, 1870: p. 248.
[122] Dupré (1807–1809), 1819: 2: p. 375. Dupré also mentions Kezzaz (unidentified) and refers to the use of alhagi manna in confectionery (*guezinguèbin*, also used to describe sweetmeats prepared from other mannas).
[123] Moghadam, 1930: p. 81.
[124] Moghadam, 1930: p. 81. There are references to Persia generally in Délile, 1812: p. 10; Ouseley, 1819–1825: 1: p. 453; Royle, 1839: 1: p. 194; Honigberger, 1852: 2: pp. 305–306; Kiepert (Haussknecht), 1868: p. 473; Baillon, 1871–1892: 1: p. 108; Bamber, 1916: p. 79; Kirtikar and Basu, 1918: 1: pp. 421–422.
[125] Vambéry, 1868: p. 242.

Tar-angubīn was collected in Khorāsān (eastern Persia) for local consumption and for export from a very early period, and it appears to have maintained its trading position more successfully than other producing regions.[126] J. E. T. Aitchison (1886) observed that the area around Rui-Khauf (Khvāf) was famous for *tarajabin*.[127] Torbat-e-Heydarīyeh was also a collecting centre.[128] In Afghanistan, supplies were obtained from the country around Herāt and Kandahar,[129] whence in the 1870's about 2000 pounds were exported to India annually.[130] This passed through Kabūl where, apparently, the product was not found locally.[131] In northern Baluchistān, "*A. camelorum* .... in certain seasons .... yields a manna."[132] Whether this was ever exploited is unknown. No doubt the main centres of production and trade have been identified, but local use over wider areas may well have passed unnoticed or unrecorded.

B. CALOTROPIS sp.

Two species, *gigantea* and *procera*, of the shrub *Calotropis* together have a very extensive distribution in the Old World, from the shores of the Atlantic to southern China. Manna is chiefly, if not exclusively, a product of *C. procera* R. Br. (Fig. 4) which dominates in the drier, western half of the zone (Map 4).[133] Otherwise the two species are apparently used interchangeably, that is according to local availability, for a wide variety of purposes. *Calotropis* features in early Hindu and Arab mythology,[134] and almost every part of the plant (leaves, flowers, roots, and particularly the acrid juice or latex) is employed in Indo-Arab and, to a lesser extent, in Chinese folk medicine.[135] Al-Samarqandī (ca. 1200) refers to *Calotropis* sp. in his medical formulary (*aqrābādhīn*) under the name *khark* (Sanskrit *arka, alarka,*

---

[126] Moghadam, 1930: p. 81.

[127] Aitchison, 1886–1887: p. 467.

[128] Sykes, 1906: p. 433. Khorāsān is mentioned by Polak, 1865: p. 286; Watt, 1889–1893: 1: p. 165; Collin, 1890: p. 103; Dymock, 1890–1893: 1: p. 419.

[129] Roxburgh, 1820–1832: 3: p. 344; Irwin, 1839–1840: p. 892; Griffith, 1847: p. 358; Brandis, 1874: p. 144; Bamber, 1916: p. 79 (Afghanistan).

[130] Flückiger and Hanbury, 1879: p. 414.

[131] Burnes, 1834: 2: p. 167.

[132] Aitchison, 1888–1894: p. 3.

[133] *C. gigantea* R. Br. is common in many parts of India, at least as far west as Baluchistān (Bamber, 1916: p. 60). It is the dominant species in Assam, Burma, Malaya, Singapore, and southern China (Hai-Nan, Yunnan) and probably other parts of South East Asia (Sprengel, 1825–1828: 1: p. 850; J. D. Hooker, 1872–1897: 4: pp. 17–18; Forbes and Hemsley, 1886–1905: 26: p. 102; Cooke, 1903–1908: 2: pp. 151–152; Gamble, 1915–1936: 2: p. 832). Aiton (1810–1813: 2: pp. 78–79) states that *C. procera* is a native of Persia, *C. gigantea* of the East Indies. Forskål (1775a: p. CVIII, no. 184) refers to *Asclepias gigantea* (*öschar*) in "Arabiae Felicis"; re-identified by Löw (1881: p. 428) and Schweinfurth (1912: p. 132) as *C. procera*.

[134] Dymock, 1890–1893: 2: pp. 428–429; Blatter, 1914–1916: pp. 244–245. See also Crowfoot and Baldensperger, 1932: p. 58.

[135] Dymock (1890–1893: 2: pp. 428–437) has an extended account. See also Fée, 1825: p. 295; Ainsley, 1826: 2: pp. 486–489; Dierbach, 1826: p. 28; Merat and Lens, 1829–1834: 1: pp. 466–467; Royle, 1839: 1: p. 275; O'Shaughnessy, 1842: pp. 452–453; Waring, 1874: p. 80; Watt, 1889–1893: 2: pp. 46–47; Achart, 1905: pp. 116–117; Kirtikar and Basu, 1918: 2: pp. 810–811; Chopra, 1933: p. 470.

FRVCTVS

Fig. 4. *Calotropis procera* R. Br. (Alpinus, 1592: pp. 36–37).

Persian *khak*).[136] Prosper Alpinus (1592) mentions the therapeutic properties of the *lacte* in his account of the plants of Egypt.[137] The *Tuḥfat
al-Aḥbāb*, a Moroccan *materia medica*, probably of the 17th century,

---

[136] Al-Samarqandī, 1967: pp. 109, 218 n., 355.
[137] Alpinus, 1592: pp. 36–37 (illustration reproduced here as Fig. 4. *Beidelsar* or *beid el ossar*
(text), the "fruit" of the plant, *ʿuschār*.) I have found no reference in his *De Medicina
Aegyptiorum* (1591).

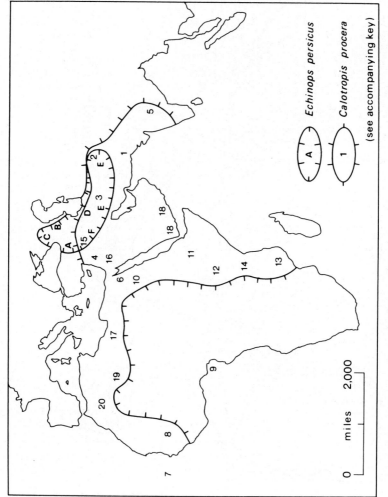

Map 4. Approximate distribution of *Echinops persicus* Stev. et Fisch. (A–F) and of *Calotropis procera* R. Br. (1–20).

Distribution of *Echinops persicus* [*E. pungens* var. *pungens*] Stev. et Fisch. ex Fisch. (Map 4).
Letters have no significance other than identification.

A: Eastern Anatolia

Tschihatchcheff, 1853–1869: 3, 2: p. 300; Hooper and Field, 1937: p. 115; Davis, 1965–1978: 5: p. 614.

B: Armenia, Transcaucasia

Tschihatchcheff, 1853–1869: 3, 2: p. 300; Davis, 1965–1978: 5: p. 614.

C: Caucasus

Ledebour, 1842–1853: 2: p. 656; Tschihatchcheff, 1853–1869: 3, 2: p. 300; Boissier, 1867–1888: 3: p. 440; Gilliat-Smith and Turril, 1930: 9: p. 433; Komarov, 1934–1962: 27: p. 33; Hooper and Field, 1937: p. 115.

D: Northern and Western Persia

Boissier, 1867–1888: 3: p. 440; Bornmüller [Strauss] 1906–1907: 20, 2: p. 157; Moghadam, 1930: pp. 46–48: Hooper and Field, 1937: p. 115; Davis, 1965–1978: 5: p. 614.

E: Central and Eastern Persia, Afghanistan

Moghadam, 1930: pp. 46–48; Gilliat-Smith and Turril, 1930: 9: p. 433; Hooper and Field, 1937: p. 115.

F: Northern Syria and Northern Iraq

Bourlier, 1857: p. 38; Moghadam, 1930: p. 48.

Distribution of *Calotropis procera* R. Br. (Map 4). Numbers have no significance other than identification.

1. Baluchistān, Sindh

J. D. Hooker, 1872–1897: 4: pp. 17–18; Burkill, 1909: p. 49; Sabeti, 1966: no. 168.

2. North West India, Afghanistan

Bellew, 1864: app. IV; Boissier, 1867–1888: 4: p. 57; J. D. Hooker, 1872–1897: 4: pp. 17–18; Bamber, 1916: p. 60; Blatter, 1919–1923: p. 295; Kitamura, 1960: p. 307; Rechinger, 1970: p. 7.

3. Persia

Royle, 1839: 1: p. 275; Boissier, 1867–1888: 4: p. 57; Dymock, 1890–1893: 2: p. 428; Cooke, 1903–1908: 2: pp. 151–152; Rechinger, 1970: p. 7.

4. Palestine

Oliver *et al.*, 1868–1937: 4: pp. 294–295; Dinsmore and Dalman, 1911: p. 168; Muschler, 1912: 2: p. 750; Temple, 1929: p. 52; Post, 1932–1933: 2: p. 192; Löw, 1967: 1: pp. 281–284.

5. Western and Central India

Boissier, 1867–1888: 4: p. 57; J. D. Hooker, 1872–1897: 4: pp. 17–18; Dymock, 1890–1893: 2: p. 428; Cooke, 1903–1908: 2: pp. 151–152; Gamble, 1915–1936: 2: p. 832; Kirtikar and Basu, 1918: 2: pp. 810–811; Blatter and Hallberg, 1919: p. 539.

6. Egypt

Sprengel, 1825–1828: 1: p. 850; Visiani, 1836: p. 15; Boissier, 1867–1888: 4: p. 57; Oliver *et al.*, 1868–1937: 4: pp. 294–295; Comes, 1879: p. 9; Woenig 1886: p. 348; Muschler, 1912: 2: p. 750; Schweinfurth, 1912: p. 10; Blatter, 1919–1933: p. 295.

7. Cape Verde Islands

Oliver *et al.*, 1868–1937: 4: pp. 294–295; Pickering, 1879: p. 330.

8. Senegal, Mauritania, Upper Guinea

Boissier, 1867–1888: 4: p. 57; Oliver *et al.*, 1868–1937: 4: pp. 294–295: Hernández-Pacheco, 1949: p. 773.

9. Fernando Po

Oliver *et al.*, 1868–1937: 4: pp. 294–295.

10. Nubia

Boissier, 1867–1888: 4: p. 57; Oliver *et al.*, 1868–1937: 4: pp. 294–295.

11. Abyssinia

Boissier, 1867–1888: 4: p. 57; Oliver *et al.*, 1868–1937: 4: pp. 294–295.

12. Uganda

Oliver *et al.*, 1868–1937: 4: pp. 294–295.

13. Mozambique

Oliver *et al.*, 1868–1937: 4: pp. 294–295.

14. Tanzania

Oliver *et al.*, 1868–1937: 4: pp. 294–295.

15. Syria

Oliver *et al.*, 1868–1937: 4: pp. 294–295; Post, 1932–1933: 2: p. 192.

(Contd. on p. 30)

16. Arabia Petraea, Sinai	Boissier, 1867–1888: 4: p. 57; Hart, 1855: p. 436; 1891: p. 99; Blatter, 1919–1933: p. 295; Post, 1932–1933: 2: p. 192.
17. Cyrenaica	Pampanini, 1914: p. 190; Trotter, 1915: p. 255; Hernández-Pacheco, 1949: p. 773.
18. Ḥaḍramawt, Aden, Jiddah	Blatter, 1919–1933: p. 295; Meulen and Wissmann, 1932: p. 158.
19. Algeria	Duveyrier, 1864: p. 180; Munby, 1866: p. 23; Boissier, 1867–1888: 4: p. 57; Hernández-Pacheco, 1949: p. 773.
20. Southern Morocco	Renaud and Colin, 1934: pp. 102, 138; Hernández-Pacheco, 1949: p. 773.

includes *'uššar* (*C. procera*);[138] and the products of the shrub could be purchased in the native pharmacies of Cairo until at least the early part of the present century.[139] Non-medicinal derivatives include dye and tannin from the bark, charcoal (for gunpowder) from the wood, fibre from the stems, and tinder and stuffing for cushions from the silky coma.[140] More locally, the juice is (or was) an ingredient in intoxicating liquor.[141]

The wide distribution of *Calotropis* spp. has naturally produced a variety of local names. The chief are *ak*, *a'k*, *mudár* (Hindi) and *'uschār* (Arabic).[142] The latter may be of Persian origin (*ushr*) and a generic name for "milk-yielding plants,"[143] later restricted in Arabic to *Calotropis* (notably *procera*). *'Uschār*, combined with the word for "sugar," provides the earliest evidence for *Calotropis* manna. Probably the first writer to describe the plant and the "sugar" was Abū Ḥanīfa al-Dīnawarī (died 895),[144] followed by Ibn Sīna (*zuccaro alhusar*),[145] Ibn Sarābī (*zuccarum haoscer*),[146] Maimonides (*sukkar al-'uṣar*),[147] and Ibn al-Baiṭār (*sokker el-o'char*).[148] European students of *materia medica* in the 16th century, such as Pietro Andrea Matthiolus, could do little more than refer to the Arab authorities, particularly Ibn Sīna and Ibn Sarābī.[149] In the *Pharmacopoea persica* of Fr. Angelus (1681) it is claimed that *schakar el ma-ascher*

---

[138] Renaud and Colin, 1934: p. 138, no. 313; cf. p. 102, no. 227.
[139] Meyerhof, 1918: p. 196.
[140] In addition to the authorities already quoted, see Bellew, 1864: Appendix IV; Burkill, 1909: p. 49.
[141] Watt, 1889–1893: 2: p. 47; Dastur, 1962: pp. 46–47.
[142] The transliteration is very variable (*'usr*, *'ušar*, *'ošer*, *'ošr*, *osciur* . . . .). See the discussion in Clément-Mullet, 1870: pp. 60–61. The plant is not certainly identifiable in the Greek and Latin authors. Royle (1837: p. 107; 1839: 1: p. 275) suggested *C. gigantea* = Greek *jumakioos*, and possibly Latin *occhi* (Pliny).
[143] Dymock, 1890–1893: 2: p. 430; Blatter, 1914–1916: pp. 244–245.
[144] Al-Dīnawarī, 1953: p. 44; Dymock, 1890–1893: 2: p. 430; Renaud, 1935: p. 66.
[145] Ibn Sīna (980–1037), 1608: 1: p. 413.
[146] Ibn Sarābī (? 12th century), 1531: p. 62; Guignes, 1905: 5: p. 532, no. 237; 6: p. 95, no. 541.
[147] Maimonides (1135–1204), 1940: p. 88, no. 178.
[148] Ibn al-Baiṭār (1197–1248), 1877–1883: 2, 1: pp. 266–269, 448–449; 3: p. 330, quoting al-Ghāfiqi (12th century). Sontheimer (1840–1842: 2: p. 36, *sukkar eluscher*) identified the plant as *C. gigantea*; according to Sickenberger (1890: p. 10) *o'char* = *C. procera*.
[149] Matthiolus, 1558: p. 245. Cf. Salmasius, 1689: 2: p. 253 (*manna alhussar*); Bochart, 1692: 3: col. 872 (*saccharum alkusar*).

(Arabic) was known as *schakar tigal* in Persia.[150] The latter was, however, *tréhala* manna, produced on *Echinops persicus* (*infra* pp. 40–42).[151]

Information on the nature, distribution and use of *Calotropis* manna is in every respect unsatisfactory. Like other mannas, it was sometimes regarded locally as a "dew," falling at night.[152] Medieval Arab and modern writers generally describe it as an "exudation" on the branches, flowers or leaves of the plant.[153] W. Dymock came to the conclusion that it was "nothing more than an exudation of the juices of the plant which naturally contain some sugar" (the latex, however, is never described as sweet). Since at least the beginning of the 19th century several authorities have expressed the view that the exudation was the result of punctures by an insect, named by J. F. Royle *gultigal* (Persian, *goltighal*).[154] This suggests further confusion with the insect product *tréhala* (*shakar tīqāl*). Whatever the mode of origin, the substance has been chiefly reported from Persia, India (*ak-* or *mudár-ka-shakar*) and southern Arabia.[155] It may also have been valued as a drug in parts of North Africa.[156]

C. COTONEASTER sp. AND ATRAPHAXIS sp.

The manna known as *schir-khecht* exudes from the branches of two shrubs, *Cotoneaster nummularia* Fischer et Meyer (1835)[157] and *Atraphaxis spinosa* L. (1753).[158] Their respective distributions are shown in Map 5. *Atraphaxis spinosa* (Fig. 5) grows to a height of about one metre and is found in dry and stony districts. *Cotoneaster nummularia* (Fig. 6), "a tall shrub or small tree" (2.5 to 5 metres), requires somewhat greater moisture. It is known as

---

[150] Angelus, 1681: pp. 361–363; followed by Virey, 1818: p. 126; Balfour, 1885: 2: p. 852; Ebert, 1908: pp. 459–460; Moghadam, 1930: pp. 41, 43, 44.

[151] According to Moghadam (1930: p. 46), *E. persicus* = *achār* (cf. Forskål, 1775a: p. LXXIII, no. 423 [*chasjīr*, *Echinops* sp.]; Schweinfurth, 1912: p. 19 ['*ichschīr*, *E. spinosus*, *E. galalensis*]). Moghadam treats *C. procera* as a case of misidentification. Note that *Apocynum syriacum* = *C. procera* (Woodson, 1930: p. 148); the former is used by Merat and Lens, 1829–1834: 4: p. 227; Schlimmer, 1874: p. 360 (*chèkkèr ol ochre, chekker rolochre*), followed by Laufer, 1919: p. 349 (*šiker al-ošr*, imported from the Yemen and the Hijaz); Collin, 1890: p. 104 (*chekerre-el-ochre*).

[152] Dymock, 1890–1893: 2: p. 430. Cf. Ibn Sīna, 1608: 1: p. 413 ("manna cadens su per alhusar . . . .").

[153] Ibn al-Baiṭār, 1877–1883: 2, 1: pp. 448–449, no. 1544; Virey, 1818: p. 126; Dymock, 1890–1893: 2: p. 430.

[154] Royle, 1839: 1: p. 275. See also Délile, 1812: pp. 9–10 (Persia, but not Egypt); Virey, 1818: p. 126; Anon., 1828: p. 262; Endlicher, 1841: p. 300; Hanbury, 1859: p. 180. Watt (1889–1893: 2: p. 47) names the insect *Larinus ursus*.

[155] Schlimmer, 1874: p. 360 (under *Apocynum syriacum*). Unidentified "manna" from southern Arabia is mentioned by Linschoten (1583–1592), 1885: 1: p. 44 (Arabia Felix), and Ovington (1699), 1929: p. 245 (Muscat).

[156] Ibn al-Baiṭār, 1877–1883: 2: pp. 448–449, no. 1544. Cf. Trotter, 1915: p. 257.

[157] Sabeti (1966: no. 264) describes *C. nummularia* as a variety of *C. racemiflorus*. In Guest and Townsend (1966–1974: 2: p. 104) *nummularia* and *racemiflorus* are treated as synonyms. Kitamura (1960: p. 173) gives *C. racemiflorus* Desv., var. *suavis* Pojark.

[158] Varieties *sinaica* (Post, 1932–1933: 2: p. 468; Sabeti, 1966: no. 128) and *rotundifolia* (Sabeti, *loc. cit.*).

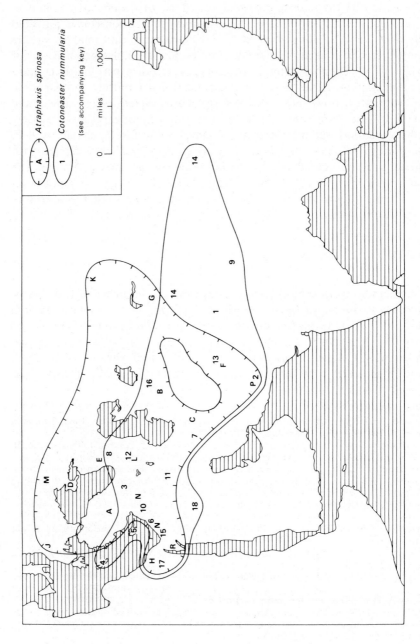

Map. 5 Approximate distribution of *Atraphaxis spinosa* L. (A–R) and of *Cotoneaster nummularia* Fisch. et Meyer (1–18).

Distribution of *Atraphaxis spinosa* L. (Map 5). Letters have no significance other than identification.

A: Asia Minor Tschihatchcheff, 1853–1869: 3, 2: p. 439; Kitamura, 1960: p. 173; Davis, 1965-1978: 2: p. 266.

B: Turkmen, Transcaspia Tschihatchcheff, 1853–1869: 3, 2: p. 439; Boissier, 1867–1888: 4: pp. 1020–1021; Paulsen, 1912: pp. 192–193; Davis, 1965–1978: 2: p. 266; Rechinger, 1968; pp. 30–31.

C: Persia L'Héritier de Brutelle, 1784: 1: no. 14; Tschihatchcheff, 1853–1869: 3, 2: p. 439; Boissier, 1867–1888: 4: pp. 1020–1021; Kiepert [Haussknecht], 1868: p. 473; Dragendorff, 1898: p. 194; Kitamura, 1960: p. 88; Davis, 1965–1978: 2: p. 266; Sabeti, 1966: no. 128; Rechinger, 1968: pp. 30–31.

D: Crimea Steven, 1856–1857: 29, 2: p. 378; Gilliat-Smith and Turril, 1930: 10: p. 481.

E: Caucasus Tschihatchcheff, 1853–1869: 3, 2: p. 439; Gilliat-Smith and Turril, 1930: 10: p. 481; Kitamura, 1960: p. 88.

F: Afghanistan Gilliat-Smith and Turril, 1930: 10: p. 481; Kitamura, 1960: p. 88; Rechinger, 1968: pp. 30–31.

G: Central Asia Boissier, 1867–1888: 4: pp. 1020–1021; Gilliat-Smith and Turril, 1930: 10: p. 481; Kitamura, 1960: p. 88.

H: Egypt Schweinfurth, 1912: p. 8; Gilliat-Smith and Turril, 1930: 10: p. 481.

J: Balkans Kitamura, 1960: p. 88.

K: Western Siberia L'Héritier de Brutelle, 1784: 1: no. 14; Tschihatchcheff, 1853–1869: 3, 2: p. 439; Kitamura, 1960: p. 88.

L: Armenia L'Héritier de Brutelle, 1784: 1: no. 14; Tschihatchcheff, 1853–1869: 3, 2: p. 439; Boissier, 1867–1888: 4: pp. 1020–1021.

M: Eastern Europe Tschihatchcheff, 1853–1869: 3, 2: p. 439; Kitamura, 1960: p. 88.

N: Syria, Palestine Boissier, 1867–1888: 4: pp. 1020–1021; Muschler, 1912: 1: p. 262; Post, 1932–1933: 2: p. 468.

P: Baluchistān Boissier, 1867–1888: 4: pp. 1020–1021.

R: Sinai Fresenius, 1934: pp. 74–75; Boissier, 1867–1888: 4: pp. 1020–1021.

Distribution of *Cotoneaster nummularia* Fisch. et Meyer (Map 5). Numbers have no significance other than identification.

1. Northern India J. D. Hooker, 1872–1897: 2: p. 386; Dragendorff, 1898: p. 273; Bamber, 1916: p. 82; Gilliat-Smith and Turril, 1930: 8: p. 386; Guest and Townsend, 1966–1974: 2: p. 104.

2. Baluchistān Boissier, 1867–1888: 2: p. 666; Burkill, 1909: p. 32; Bamber, 1916: p. 82; Guest and Townsend, 1966–1974: 2: p. 104.

3. Eastern and Central Anatolia Tschihatchcheff, 1853–1869: 3, 1: p. 128; Boissier, 1867–1888: 2: p. 66; J. D. Hooker, 1872–1897: 2: p. 386; Gilliat-Smith and Turril, 1930: 8: p. 386; Zohary, 1950: p. 78; Davis, 1965–1978: 4: pp. 131–132; Guest and Townsend, 1966–1974: 2: p. 104.

4. Crete Davis, 1965–1978: 4: pp. 131–132.

5. Cyprus Davis, 1965–1978: 4: pp. 131–132.

6. Lebanon Boissier, 1867–1888: 2: p. 666; Davis, 1965–1978: 4: pp. 131–132.

7. Persia Boissier, 1867–1888: 2: p. 666; Aitchison, 1886–1887: p. 467; Dragendorff, 1898: p. 273; Gilliat-Smith and Turril, 1930: 8: p. 386; Hooper and Field,

(Contd. on p. 34)

	1937: p. 106; Davis, 1965–1978: 4: pp. 131–132; Sabeti, 1966: no. 264; Guest and Townsend, 1966–1974: 2: p. 104.
8. Caucasus	Ledebour, 1842–1853: 2: p. 93; Tschihatchcheff, 1853–1869: 3, 1: p. 128; Boissier, 1867–1888: 2: p. 666; Gilliat-Smith and Turril, 1930: 8: pp. 386–387; Davis, 1965–1978: 4: pp. 131–132; Guest and Townsend, 1966–1974: 2: p. 104.
9. Tibet	J. D. Hooker, 1872–1897: 2: p. 386; Franchet, 1883–1888: 8: p. 225.
10. Syria	Boissier, 1867–1888: 2: p. 666; Gilliat-Smith and Turril, 1930: 8: p. 386; Davis, 1965–1978: 4: pp. 131–132.
11. Iraq	Boissier, 1867–1888: 2: p. 666; Gilliat-Smith and Turril, 1930: 8: p. 386; Zohary, 1950: p. 78; Guest and Townsend, 1966–1974: 2: p. 104.
12. Transcaucasia	Tschihatchcheff, 1853–1869: 3, 1: p. 128; Boissier, 1867–1888: 2: p. 666; Gilliat-Smith and Turril, 1930: 8: p. 386; Zohary, 1950: p. 78; Kitamura, 1960: p. 173.
13. Afghanistan	J. D. Hooker, 1872–1897: 2: p. 386; Aitchison, 1888–1894: p. 22; Kitamura, 1960: p. 173; Guest and Townsend, 1966–1974: 2: p. 104.
14. Western and Central Asia	Boissier, 1867–1888: 2: p. 666; J. D. Hooker, 1872–1897: 2: p. 386; Lansdell, 1885: 2: p. 633; Forbes and Hemsley, 1886–1905: 36: p. 470; Gilliat-Smith and Turril, 1930: 8: p. 386; Zohary, 1950: p. 78; Kitamura, 1960: p. 173; Diels, 1961: p. 385; Guest and Townsend, 1966–1974: 2: p. 104.
15. Palestine	Boissier, 1867–1888: 2: p. 666; Temple, 1929: p. 64; Post, 1932–1933: 1: p. 458.
16. Turkmen, Transcaspia	Guest and Townsend, 1966–1974: 2: p. 104.
17. Egypt	Dragendorff, 1898: p. 273; Gilliat-Smith and Turril, 1930: 8: p. 386; Post, 1932–1933: I: p. 458.
18. Arabia	Philby, 1922: 2: p. 312.

siah-chob, "black wood" or "black stick," in north-western India and Afghanistan, and as *kashiru* in Persia.[159]

The manna has only been reported from (a) the neighbourhood of Tehrān, including the foothills of the Elburz mountains,[160] (b) Khorāsān (Persia), extending eastward to the Paropamisus range (Afghanistan),[161] and (c) parts of Turkmen/Turkestan (Map 6). Whether this is due to limited recognition of the product[162] (unlikely in view of its reputation), or to the presence of particular varieties of the shrubs, or again to the localized effect of some parasitic insect, appears not to have been established.

Schir-khecht "occurs in small yellowish-white granules, about the size of millet seed."[163] "During July .... the smaller branches of the *cotoneaster*

---

[159] Aitchison, 1886–1887: p. 467; 1888–1894: p. 64; Dymock, 1890–1893: 1: p. 583; Dragendorff, 1895: p. 273.

[160] Polak, 1865: 2: p. 286; Moghadam, 1930: p. 96 (sample from Kân, 15 kilometres north west of Tehrān); both with reference to *Atraphaxis spinosa*.

[161] Aitchison, 1888–1894: p. 22; Dymock, 1890–1893: 1: p. 583; Moghadam, 1930: p. 96; all with reference to *Cotoneaster nummularia*.

[162] Cf. Guest and Townsend, 1966–1974: 2: p. 104 ("There appears to be neither any general Kurdish name nor any locally recognised use for [*C. nummularia/racemiflorus*].")

[163] Dymock, 1890–1893: 1: p. 585 (with a chemical analysis of *Cotoneaster* manna). Cf. descriptions in Flückiger and Hanbury, 1879: p. 415; Ducros, 1930: p. 128. Ludwig (1870: pp. 47–48) gives the chemical composition of *Atraphaxis* manna.

34

Fig. 5. *Atraphaxis spinosa* L. (Dillenius, 1732:1: tab. XL.).

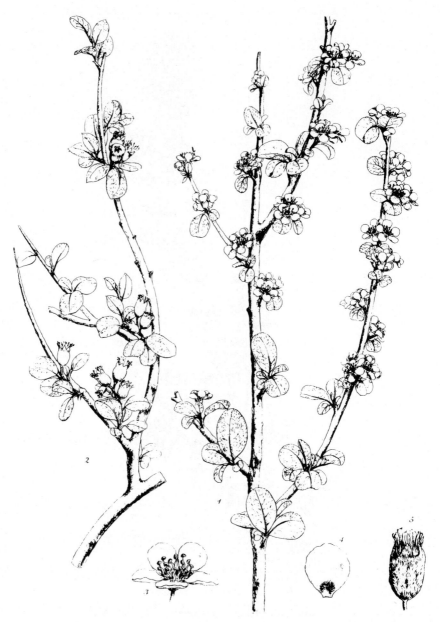

Fig. 6. *Cotoneaster nummularia* Fisch. et Meyer (Aitchison, 1888–1894: pl. 9).

become covered with the exudation, and this is collected by merely shaking the branches over a cloth."[164] It is valued chiefly as a medicinal product (pectoral, purgative) and more particularly as a remedy for typhoid.[165] S. Moghadam observed that it was imported into Persia from Herāt at times of epidemic.[166] Like other mannas of the same region, however, it is sometimes incorporated in sweetmeats (? of therapeutic reputation).[167] The substance is (or was) marketed in a more or less pure state and also mixed with flour.[168] Apparently from the early medieval period schir-khecht was available in pharmacies between Cairo[169] and northern India, where it was more highly regarded than the manna of Alhagi maurorum.

Schir-khecht (Persian) means "dried (or hardened) juice (or milk)." Moghadam gives two variants: shir-khocht, "latex desséché," and schiré khachak, "latex de l'arbrisseau épineux"[170] (presumably Atraphaxis spinosa). Garcia da Orta (1563) has xirquest or xircast, "milk [xir, shir, schir] of a tree called quest."[171] This is followed by Carolus Clusius (ca. 1600)[172] and Pedro Teixiera (1586–1605).[173] In India, shirkhisht, or some cognate word, apparently came to be the generic name for a variety of imported mannas.[174] The mann fārsī or "Persian manna" of Cairo included that of Astragalus and Alhagi as well as Cotoneaster and Atraphaxis.[175]

Arabo-Persian authors from the 10th to the 13th century are the first to refer to schir-khecht. It is mentioned as one of the products (shīrkhisht) of Herāt in the anonymous Persian geography Ḥudūd al-Ālam of A.D. 982.[176] The physicians Muwaffiq ibn 'Alī of Herāt (ca. 970),[177] Ibn Sīna (980–1037),[178] Ibn Buṭlān (ca. 1050),[179] and al-Samaqandī, who died in Herāt in

---

[164] Aitchison, 1886–1887: p. 467.
[165] Hooper and Field, 1937: p. 106.
[166] Moghadam, 1930: p. 96.
[167] Aitchison, 1886–1887: p. 467; Hooper and Field, 1937: p. 107.
[168] Schlimmer, 1874: pp. 357–358; Dymock, 1890–1893: 1: p. 584; Moghadam, 1930: p. 96; Leclerc in Ibn al-Baiṭār, 1877–1883: 2: p. 358, quoting Dā'ūd al-Anṭākī (ca. 1590).
[169] Ducros, 1930: p. 128; Meyerhof in Maimonides, 1940: p. 194, no. 386. Cf. Alpinus, 1591: p. 127; Leclerc in Ibn al-Baiṭār, 1877–1883: 2: p. 358, quoting Dā'ūd al-Anṭākī (ca. 1590).
[170] Moghadam, 1930: p. 95.
[171] Orta, 1913: pp. 280–281. Cf. Leclerc in Ibn al-Baiṭār, 1887–1883: 2: p. 358. Steingass (1957) gives shīr-khisht, "manna [of Atraphaxis spinosa]," and shīr-khushk, "a kind of manna." khushk (xušk, *xišk) = "dry"; (šīr)-xušt, and (šīr)-xišt = "dried up" (more appropriate). According to Sinclair and Ferguson in Teixeira, 1902: p. 203, n. 4, khisht = Cotoneaster nummularia. Cf. Wulff, 1966: pp. 78, 302, šīrḫešk = C. nummularia; similarly Gauba, 1949–1953: 58: p. 123 (schirchesht). As in the case of the willow, the name of the manna and of the tree or shrub appear to be virtually interchangeable.
[172] Clusius, 1605: p. 164. Clusius (Charles de l'Ecluse, 1525–1609) was director of the botanical garden of Leiden university.
[173] Teixeira, 1902: p. 203; also J. Bauhin, 1650–1651: 1: p. 199 (arbore quest). "Milk of a tree" in Acosta, 1578: p. 308; Linschoten (1583–1592), 1885: 2: p. 100; G. Bauhin, 1671: p. 496; Pétis de la Croix, 1722: pp. 204–205.
[174] Dymock, 1890–1893: 1: p. 584; Watt, 1889–1893: 5: p. 166; Birdwood and Foster, 1893: p. 287 n. 3.
[175] Ducros, 1930: p. 128, no. 223; Meyerhof in Maimonides, 1940: p. 194, no. 386.
[176] Anon., 1937: p. 104. The author's chief source appears to have been Ibrāhīm al-Fārisī al-Iṣṭakhrī (ca. 950).
[177] Muwaffiq ibn 'Alī, 1968: pp. 173, 355 (schir-chischt).
[178] Ibn Sīna, 1608: 1: p. 359 (asceheracost; siracost in the margin).
[179] Ibn Buṭlān, 1531: p. 24 ("Sirusuk est ros qui cadit in Chora [Khorāsān] . . . .").

37

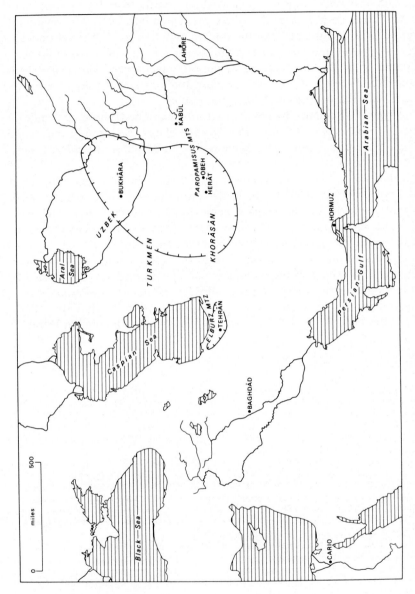

Map 6. Approximate limits of areas producing and marketing *schir-ḵhecht* (on *Atraphaxis spinosa* L. and *Cotoneaster nummularia* Fisch. et Meyer).

38

1222,[180] include *schir-khecht* among other mannas. Ibn al-Baiṭār (1197–1248) devotes a section to *chīr khocht* (*siracost*), affirming that it is a manna of high repute that "falls" upon willows (confusion with *bīd-khecht* of *Salix fragilis*) in the vincinty of Herāt.[181]

It is not known whether *schir-khecht* was available anywhere in Europe at this time – possibly, however, in Palermo and in the cities of southern Spain. The only European who may refer to it (or, more likely perhaps, the manna of *Alhagi* spp.) is Jordanus (ca. 1320–1330), a French Dominican and for a brief period bishop of Columbum in Malabar. In his *Mirabilia Descripta* he observed that "between this country of Persia and India the Less (? Sindh, south and east to the borders of Malabar) is a certain region (? Khorāsān) where manna falls in a very great quantity, white as snow, sweeter than all other sweet things, delicious, and of an admirable and incredible efficacy."[182]

Several European travellers of the second half of the sixteenth and the early seventeeth centuries refer to trade in *schir-khecht*. The original statement (first published in 1563 and much copied) is by Garcia da Orta, a Portuguese botanist who went to India as physician to the viceroy. "I only know," he wrote, "that [manna] is brought in three ways to Ormuz [Hormuz, at the mouth of the Persian Gulf] from the province of the Uzbegs. The chief kind is . . . . what you see in the druggists' shops in flasks, like sweetmeats, and with the taste of honeycomb. They call it *xirquest* or *xircast* . . . . it is the dew that falls from these trees, or the gum that exudes from them. We have corrupted the word to *siracost*. Avicenna [Ibn Sīna] belonged to that country of the Uzbegs, native of a city called Bocora [Bukhāra] . . . . so that it is natural that he should know the name very well. . . ."[183] Several centuries later, the Russian traveller N. Murav'ev (1819–1820) found that the Uzbeks were "very fond of sweetmeats, especially sugar and comfits."[184]

Christovão Acosta's source of information seems to have been exclusively Orta's *Colloquies*.[185] Nor does the Dutchman Huyghen van Linschoten, who visited Goa in the course of his travels in the East Indies (1583–1592) add much to our knowledge,[186] but we gather, elsewhere in the narrative, that the island of Hormuz was a great entrepôt for goods from Arabia, Persia,

---

[180] Al-Samarqandī, 1967: pp. 72, 186 n. 120 (*siracost*).
[181] Ibn al-Baiṭār, 1877–1883: 2: pp. 357–358, no. 1380; cf. Sontheimer, 1840–1842: 2: p. 118 (*schir chaschak*). Earlier, Ibn Buṭlān associated this "dew" (*ros*) with the willow; also Dā'ūd al-Anṭākī (died 1599), quoted by Leclerc in Ibn al-Baiṭār, 1877–1883: 2: p. 358 ("une rosée qui tombe sur les arbres, notamment sur le *khiláf* [willow].")
[182] Jordanus, 1863: p. 10.
[183] Orta, 1913: p. 281. According to Pétis de la Croix (1722: pp. 204–205), "[Uzbek] merchants trade chiefly in the higher Tartary, Muscovy, and the Indies. They carry thither silks and a great deal of manna (*scherkest*) . . . ."
[184] Murav'ev, 1871: p. 161.
[185] Acosta, 1585: p. 308. Acosta, a Portuguese naturalist and physician, visited India to collect information for his *Tractado de las Drogas y Medicinas de las Indias Orientalis*, first published in Burgos, 1578.
[186] Linschoten, 1885: 1: p. 100.

Turkestan and India.[187] Another Portuguese, Pedro Teixeira, who spent several years in Persia (ca. 1600), tells us that the "city of Hrey (Rey Xarear or Xaharihrey, the ancient *Rhagae*, about 10 kilometres south-south-east of Tehrān) is famous, as much for its size as for some things found therein, of which I will mention only mana (*xirquest*), as the best and purest yet known. It is taken hence in great quantity to Harmuz, and exported thence to all the East."[188] The orientalist Barthélemy d'Herbelot (1697) described *schirkiest* as "manne de Reï."[189] Teixeira also mentions another "Hrey [in Khorāsān] where is gathered much and perfect manna, and whose walls are washed by the cool river Habin" (apparently Obeh on the Hari Rūd, about 100 kilometres east of Herāt).[190]

We have then, by ca. 1600, evidence of the two main areas of *schir-khecht* production – the neighbourhood of Tehrān (from *Atraphaxis spinosa*), and Khorāsān/western Afghanistan/southern Turkestan (from *Cotoneaster nummularia*). European scholars of the 17th and 18th centuries drew largely, if not exclusively, upon these earlier accounts, including those of the Arabs.[191] *Schir-khecht* reached the bazaars of Lahore, and presumably other parts of India, from Persia and Turkestan by way of Kabūl during the 19th century.[192] And, as we have seen, it could be purchased in Cairo in the 1930's and 1940's. Doubtless the manna is still widely available, a thousand years or more after the first known reference to its use.

D. ECHINOPS spp.

The medicinal manna known as *tréhala* or *tricala* is obtained from the nidus or cocoon case of a beetle, *Larinus maculatus* Faldermann. This is found on the leaves and stalks of species of *Echinops*, chiefly *E. persicus* Steven et Fischer ex Fischer (1812)[193] (Fig. 7) and probably *E. candidus* Boissier (in central Persia).[194] The known distribution of the former extends from eastern Anatolia and Transcaucasia, through Persia to the borders of Afghanistan (Map 4).[195] "Tréhala," according to S. Moghadam, is a corrup-

---

[187] Linschoten, 1885: 1: p. 47. Hormuz was originally on the mainland. It was moved (? ca. 1300) to the north side of the island, and was most flourishing between the 14th and the 16th centuries.

[188] Teixeira, 1902: p. 203 (followed by Stevens, 1715: pp. 29–30). In 1610, Teixeira published an important *Account of the Kings of Persia and Ormuz*. Rey was the capital city of part of northern Persia (also known as Rey) under the Caliphate.

[189] Herbelot, 1777–1779: 2: p. 546. He thought that this substance was the same as the manna of Calabria.

[190] Teixeira, 1902: pp. 244 n. 3, 248 and n. 5. Altomarus (1562: pp. 10–11) refers to the *siracost* of *Corasceni in Oriente*.

[191] In addition to Clusius and G. and J. Bauhin (*supra* p. 37 nn. 172, 173), see Salmasius (1629), 1689: 2: p. 249 (*xirquest*); Deusingius, 1659: p. 9 (*schirchoscht*); Johnstone, 1662: p. 333 (*siracost, xercost*); Bochart (1599–1667), 1692: 3: p. 871 (*zirkest, zirakost*).

[192] Royle, 1839: 1: pp. 226–227; Honigberger, 1852: 2: pp. 305–306; Watt, 1890–1893: 3: p. 443. Cf. Kiepert (Haussknecht), 1868: p. 473.

[193] Fischer, 1812: p. 37 (*nomen nudum*). Now placed within *E. pungens* (var. *pungens*) Trautvett. See Ledebour, 1842–1853: 2: p. 656; Boissier, 1867–1888: 3: p. 440; Davis, 1965–1978: 5: p. 614; Komarov, 1952–1962: 27: p. 33.

[194] Boissier, 1867–1888: 3: p. 435 (Ispahan; Elburz mountains).

[195] Sprengel, 1825–1828: 3: p. 396.

Fig. 7. *Echinops persicus* Stev. et Fisch. ex Fisch. (Moghadam, 1930: pl. II).

tion of "Téhérani."[196] The Persian name is *shakar tīqāl*, "sugar of nests," first reported by Fr. Angelus (*schakar tigal*), who also refers to the insect as *c-hezoukek*.[197] M. Meyerhof has suggested that a manniparous insect mentioned by al-Bīrūnī (973–1050), in an article on the manna-producing plant

---

[196] Moghadam, 1930: pp. 41, 44. Bourlier (1857: p. 38) states that in Syria the product is known as *thrâne*, from which he derives *thrâle, trehala, tricala*.
[197] Angelus, 1681: pp. 361–363. Hanbury (1859: p. 180) gives *shek roukeh*, and Moghadam (1930: p. 44) *khozoukâk* (Shīrāz); probably also the *gultigal* of Royle, 1839: 1: p. 275.

41

al-ḥāj (*Alhagi maurorum* or *A. camelorum*), may in fact have been *L. maculatus* on *E. persicus*.[198] Angelus assumed that *schakar tigal* and *schakar el ma-ascher* (? *Calotropis* sp.) were one and the same.

Tréhala was brought to the notice of the West by J. M. Honigberger (1852) who described and illustrated the "insect nests" (*shukure teeghal*).[199] About the same time, specimens from Persia were presented to the British Museum.[200] The product was readily available in the pharmacies of Constantinople (and possibly in centres of Islamic culture further west), and it was from here that *tréhala* displayed at the Paris Exhibition of 1855 was obtained. Descriptions and chemical analyses were later published by C. Bourlier (1857), a resident of Constantinople, N. J. B. G. Guibourt (1858) and M. Berthelot (1859).[201] Guibourt named the insect *Larinus nidificans*, and Berthelot isolated and named the sugar *tréhalose*, amounting to 17.50 to 28.80 per cent of the substance as a whole.[202] The insect was subsequently found to correspond to *L. maculatus*, reported by Francisco Faldermann in his *Fauna Entomologica Trans-Caucasica* (1837).[203] According to the entomologist H. Jekel, "this species has a very extended habitat,"[204] which suggests the existence of other host plants.[205]

A decoction of *tréhala* was used to relieve respiratory ailments, and it appears to have enjoyed a considerable reputation in different parts of the Islamic world. This implies extensive trading on account of the comparatively limited area within which the substance was produced. However it was not among the alimentary mannas and the amounts involved in trade were probably small.

---

[198] Meyerhof, 1947: pp. 31–36 (p. 33: "a hollow in which is generated an animal"). Al-Bīrūnī was quoting from an earlier authority, al-Fazārī. Both host plants are spiny which may have led to an error in identification. In any event, as ponted out by Meyerhof, this is by far the earliest reference to a manniparous insect.

[199] Honigberger, 1852: 2: pp. 305–306. He states that the manna was imported from Hindustan (rather than from Persia) to Lahore. This seems to be a misunderstanding, or further confusion with the manna of *Calotropis procera*.

[200] Hanbury, 1859: p. 181.

[201] Bourlier, 1857: pp. 37–38; Guibourt, 1858a: pp. 1213–1217 (described as "cette substance alimentaire .... jusqu'ici complétement inconnue"); Berthelot, 1859: pp. 272–282. Bourlier located a sample of *tréhala* (of unstated provenance and date) in the Paris collection of the entomologist M. Chevrolat. See also Apping, 1885.

[202] Cf. Ebert, 1908: p. 460; Moghadam, 1930: p. 45. S. G. Harrison (1951: p. 409) observed that "Certain fungi have been associated with manna on the grounds that they contain .... trehalose .... Trehalose has been found, for example, in *Boletus edulis* Bull. ex Fr. (*Polyporaceae*) and in *Lactarius piperatus* (Scop. ex Fr.) Fr. (*Agaricaceae*)." Cf. Al-Bīrūnī, 1973: 1: p. 310.

[203] Faldermann, 1837: pp. 228–229, and Tab. 6 (10); Schoenherr, 1833–1838: 3: p. 112; Ludwig, 1870: pp. 49–50; Haussknecht, 1870: p. 249; Schlimmer, 1874: pp. 359–360; Ebert, 1908: pp. 459–460. Cf. Grassé, 1949–1952: 1: p. 985 ("Les coques de nymphose de divers *Larinus* Germ. [order *Coleoptera*, family *Curculionideae*], très riches en tréhalose, ont été l'objet d'un commerce très important dans le Proche-Orient.")

[204] "European Turkey (Frivaldski), Berouth, Caucasus, Persia (Dupont), etc; and it is recorded by Schönherr as also found in Barbary and Portugal." (Hanbury, 1859: p. 179). See also Jekel, 1849: pp. 151–153, no. 348.

[205] *Calotropis procera*, according to Meyerhof, 1947: p. 35. Hanbury (1859: pp. 181–183) describes, on the basis of another report by Jekel, "a second insect product [obtained 1851] .... resembling dark honey" and exuded by *E. persicus* when pierced by *L. mellificus* Jekel. This seems, however, to be no other than *tréhala* (Dragendorff, 1898: p. 684; Kaiser, 1924: pp. 114, 131).

E. LECANORA sp.

Lichens have a remarkably wide geographical range and they have been put to a variety of uses. Several are rich in colouring matter, particularly scarlet-purple (*Roccella tinctoria* C., *Lecanora tartarea* Mass., and *L. parella* Mass.). Others were incorporated in the *materia medica* of medieval Europe, North Africa and the Orient. Few if any are poisonous to man. Quantities of *Evernia* [*Parmelia*] *furfuracea* E. Fr. have been found in Egyptian tombs of the XX–XXII dynasties (ca. 1200–900 B.C.). Until recent times this lichen (imported from the Aegean) and *E. prunastri* L. (Ach.) were sold in Cairo under the name of *sheba*, as a kind of yeast or agent of fermentation.[206]

The better-known nutritive lichens[207] include *Cetraria islandica* L. ("Iceland moss"), *Cladonia rangiferina* (L.) Web. and *C. sylvatica* (L.) Hoffm. ("reindeer moss"), *Gyrophora* spp. ("tripe de roche"), and *Lecanora esculenta* Evers. The use of lichens as food is probably very old. The practice has survived into modern times chiefly in impoverished environments, such as deserts, steppes and tundra, beyond the limits of cultivation; elsewhere, lichens are occasionally reported as a famine food. *Lecanora esculenta* has been identified with the manna of the Hebrews (Exodus 16. 20–35, Numbers 11. 6–9), although the lichen has not been observed in Sinai, either *in situ* or in the form of wind-borne deposits. However this in itself is not decisive for explicit reports of such deposits elsewhere in western Asia do not go back beyond the 19th century.[208]

The "manna lichen," now generally known as *Lecanora esculenta* Eversmann (1831), has been placed within several genera: *Lichen* (Ammann, 1739; Pallas, 1771–1776; Acharius, 1803; Link, 1848), *Urceolaria* (Acharius, 1810, 1814), *Parmelia* (Sprengel, 1825–1828; Montagne, 1846), *Chlorangium* (Link, 1849; J. Müller, 1858; Pitra, 1868), and *Aspicilia* (Elenkin, 1901).[209] A. Elenkin distinguished the following forms of *Aspicilia alpino-desertorum: esculenta-alpina, affinis,*[210] *fruticulosa,*[211] and *fruticulosa-foliacea* (alpine); *desertoides, foliacea,* and *esculenta-tesquina* (steppes and deserts). The lichen collected in Spain, according to the Arab botanists, was apparently *Lecanora crassa.*[212]

---

[206] Forskål, 1775a: p. 193; Délile, 1812: p. 80 (*cheybeh*); Loret, 1887: no 134; Schweinfurth, 1912: p. 34 (*schēbe*); 1918: pp. 440–441; Meyerhof, 1918: p. 196 (*schēba*); Meyerhof and Sobhy in al-Ghāfiqi, 1932–1938: 1: pp. 64–66 (*shēba*). Acloque (1893: p. 311) observed that "les Turcs préparent leur pain avec de l'eau dans laquelle ils ont fait bouillir ce lichen [*Evernia prunastri*]; elle donne à la pâte une saveur qui leur plaît." *Evernia furfuracea* in powdered form has also been used as an absorbant in perfumery.

[207] Zopf, 1896: pp. 185–187; Abbayes, 1951: p. 86.

[208] Muḥammad al-Ṭamīmī (ca. 980) described two kinds of edible lichen; the first was "the airy, dry substance that falls on trees" (Hamarneh, 1973: p. 78), a possible reference to *L. esculenta.*

[209] Also, less frequently, *Peltidea* (Acharius, 1803), *Arthonia* (Acharius, 1806, 1810), *Placodium* (Link, 1848 [Algeria]).

[210] Cf. Eversmann, 1831: pp. 354–355.

[211] *L. esculenta* var. *fruticulosa* of Eversmann, 1831: p. 354; Nylander, 1858: p. 113, no. 23; Hue, 1891: p. 74.

[212] Meyerhof and Sobby in al-Ghāfiqi, 1932–1938: 2: p. 418, quoting Sickenberger, 1893.

Crusts of granulose *L. esculenta*[213] are found chiefly on limestone exposures or on lime-rich soil.[214] After periods of drought, the thallus easily becomes detached in the form of irregular nodules 5 millimetres to 15 millimetres in diameter; "externally these are clear brown or whitish; the interior is white and consists of branching interlaced hyphae, with masses of calcium oxalate crystals averaging about 60 per cent or more of the whole substance."[215] A. Kerner von Marilaun observed that ten fragments together weighed no more than 3.36 grams.[216] Driven before the wind, the lichens accumulate in depressions, behind sand dunes and along ravines, whence they are sometimes carried away by flash floods. Extensive deposits up to 20 centimetres deep have been reported.

(a) Hispano-Arab reports of Lecanora esculenta

According to J. Schlimmer, the inhabitants of Sīstan in eastern Persia preserve the tradition that part of the army of Alexander the Great (330–327 B.C.) was saved from starvation by recourse to the lichen *esculenta*.[217] The earliest written references, apart from a possible biblical allusion, are to be found in Arabic works of the 9th to the 13th centuries.

Lucien Leclerc in his commentary upon Ibn al-Baiṭār *Traité des simples* suggested that the substance known as *djouz djondom* (Persian) or "nut-like wheat" was *Lecanora [esculenta]*,[218] and this seems to have been generally accepted.[219] Muḥammad al-Ghāfiqi (ca. 1160), who lived in Andalusia, gives the synonyms "honey dust" and "flower of the stone" and states that the "earth" *gawz gundum* was exported from Barca (Cyrenaica) and Khorāsān (ancient Parthia).[220] Here he is quoting from Ibn Sīna (980–1037, and a native of Bukhāra), from whom we also learn that this *terra granosa* was used in various medicinal compounds and in the preparation of honey wine.[221] Al-Bakrī (ca. 1040–1094) likewise referred to the substance as "earth" (*torba*), found near Barca.[222] The great geographer and natural

---

[213] First illustrated by Pallas, 1771–1776: 4: Tab. III, 1, i, Fig. 4; 1788–1793: 5: Tab. XXI, Fig. 2. See also Treviranus, 1816: Taf. III, 20–23; Eversmann, 1831: Tab. LXXVIII.

[214] Cf. Murchison, 1864: p. 769 (granite and sandstone, as well as calcareous rocks [Asia Minor]); Henneguy, 1883: p. 105 ("les silex" [Algeria]); Flagey, 1896–1897: p. 52 ("sur la terre siliceuse et les rochers dolomitiques sahariens"); Elenkin, 190lb: pp. 35–36 ("ad terram argillosam"). Steiner (1899 p. 293, no. 17) found the lichen on volcanic tufa near Erevan, Armenia.

[215] Smith, 1921: p. 405. According to Teesdale (1897: p. 229), some Algerian varieties have a reddish appearance.

[216] Marilaun, 1894–1895: 2: pp. 810–811.

[217] Schlimmer, 1874: p. 12.

[218] Ibn al-Baiṭār, 1877–1883: 1: pp. 386–387, no. 538.

[219] Earlier, Sontheimer (1840–1842: 1: p. 374) suggested that *dschawz dschandum* was the resin of *Garcinia mangostana* L. Cf. 'Isā, 1930: p. 86, no. 10.

[220] Al-Ghāfiqi, 1932–1938: 2: pp. 418–420. Cf. Ibn Sarābī (? 12th century), 1531: p. 75, cap. LXXIX (*ieum henden*, *nux henden*); Guignes, 1905: 5: p. 541, no. 287 (*ieuz hendem*). The "manna from Parthia" mentioned by the Byzantine historian Georgius Syncellus ([ca. 800] 1652: pp. 128–129) was possibly *L. esculenta*.

[221] Ibn Sīna, 1608: 1: p. 362 (De [nuce henden]: *giauzi henden*. "Nux henden quid est? Est terra granosa, sicut cicer, alba, declinans ad citrinitatem, quae exportatur ex Barcha, et ex Corasceni, ex qua cum melle sit vinum.")

[222] Al-Bakrī 1913: p. 15 ("qui sert a faire fermenter le miel"). Elsewhere (p. 104), al-Bakrī writes of "manna" collected at Touzer (Tozeur), Tunisia.

historian al-Idrīsī (1100–1166) wrote, more fully and more accurately, "[*Gawz gundum*] is something that grows in deserts which cross the centre of sterile mountains. It grows between stones, is yellowish in colour and does not rise above the soil higher than the size of a finger-nail. Al-Rāzī [865–925], al-Basrī and Is-hāq ibn 'Imrān [of Qairwan, early 9th century] call it "dandruff of the stone" (*bahaq al-hagar,* lichen). The best kind is that which is imported from Khorāsān. It occurs also in our land, in the east of Andalusia, in the mountains around Saragossa, but it is not of the same quality as that which is imported from Khorāsān. Our people collect it when it is dry. It looks then a kind of granular earth like chickpeas and is of a greyish colour."[223] Al-Idrīsī too mentions that the powdered lichen was added to honey-water. Maimonides (1135–1204), who was born in Córdoba but worked chiefly in Cairo, referred to the "earth" as *dādī,* the name by which it was known in the Maghreb.[224] Ibn al-Baitār (born in Malaga, 1197, died in Damascus, 1248) drew upon the whole corpus of Islamic pharmaceutical literature and particularly the work of al-Ghāfiqi. *Djouz djondom,* he observed, was called "fat of the earth," "honey earth" in eastern Spain, and "pigeons' dung" at Rakkah (Raqqa) on the Euphrates.[225] According to one of his authorities, Ibn Gulgul, the product was brought from the *Zāb* of Qairwan.[226]

The medico-botanists of the golden age of Islamic science did not confuse *L. esculenta* with the saccharine substances known as "manna" (Arabic *mann*); nor do they refer to the lichen as a famine food, either alone or mixed with wheat or barley. In the western half of the Arab world particularly, its chief value seems to have been as an agent of fermentation (or preservation).

(b) Distribution

West-Central Asia and South-East Europe (Map 7)

Lecanora esculenta is a product of the Irano-Turanian steppes and deserts (west-central Asia and the highlands of North Africa). The Asiatic distribution – *in situ* or as an erratic – has been described only in the most general terms and Map 7 is no more than a first approximation.[227] Medieval Arab scholars refer to supplies from Khorāsān. Its rediscovery should probably be credited to Johann Ammann who, in his *Stirpium Rariorum in*

---

[223] Ms. quoted by Meyerhof and Sobhy in al-Ghāfiqi, 1932–1938: 2: pp. 419–420.

[224] Maimonides, 1940: p. 46, no. 86. Meyerhof (ed.) comments on the origin and meaning of the word *dādī.* Cf. Ibn Sarābī (? 12th century), 1531: p. 43, cap. XXIIII (*de dadi*); Guignes, 1905: 5: p. 511, no. 139; Meyerhof, 1944–1945b: p. 1861 (*dadhi*).

[225] Cf. II Kings 6.25 (". . . . and there was a great famine in Samaria . . . and the fourth part of a cab of dove's dung [was sold] for five pieces of silver. Discussed by Bochart, 1663: 2: p. 44 (*giauz gendem,* chickpea).

[226] Alessandro Zorzi ([1519–1524] 1958: p. 113), using information supplied by Fr. Nicholas of San Michele, Murano, wrote of "manna" gathered to the south of here. He may have been referring to the manna lichen, but more likely to the saccharine substance found on certain palms.

[227] There is a sketch map in Elenkin, 1901a: p. 26.

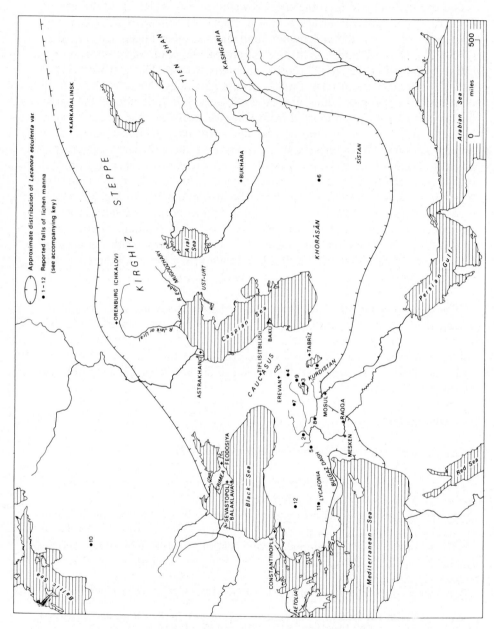

Map 7. *Lecanora esculenta* Evers., western and central Asia.

46

Reports of falls of "lichen manna," *Lecanora esculenta* Evers. (Map 7). Numbers have no significance other than identification.

1. Urmia (Reẓā'iyeh)	1829	Léveillé, 1842: p. 140; Aucher–Éloy, 1843: 2: p. 399 n.l; Hogg, 1849: pp. 223–234; 1864: p. 205; Schlimmer, 1874: p. 13 (no date); Teesdale, 1897: p. 230.
	1846	Meyer, 1847: p. 237; Maurizio, 1932: p. 99.
2. Karput (Charput, Harput)	March 1864	Reichardt [Haidinger], 1864: p. 553; Murchison, 1864: p. 769; Ludwig, 1866: pp. 284, 287 (no date); 1870: p. 52; Visiani, 1867: p. 199; Schlimmer, 1874: p. 13; Teesdale, 1897: p. 231.
3. Van	Spring 1841	Reissek, 1847: p. 195; Ludwig, 1866: p. 287 (no date); Teesdale, 1897: p. 230.
	–1864	Murchison, 1864: p. 769.
4. Ararat	[–]1828	Thénard, 1828: col. 55; Goebel, 1830: p. 393 (early 1828 [Parrot]); Reissek [Parrot], 1847: p. 196; O'Rorke [Thénard], 1860: p. 417; Schlimmer [Thénard], 1874: p. 13.
5. Malatia (Malatya)	–1864	Murchison, 1864: p. 769; Ludwig, 1866: p. 287 (no date).
6. Herāt	–1847	Decaisne, 1847: p. 314.
7. Erzurum (region)	–1857	Berkeley, 1857: p. 383.
8. Diyarbakir (Diarbékir)	–1864	Murchison, 1864: p. 769.
	[May] 1890	Anon., 1891a: p. 255; Errera, 1883: p. 84.
9. Byazid (Caldiran)	April and June 1849	Berkeley [Heinig], 1849–1864 (1849): p. 581.
10. Smorgonie (Smorgon')	March 1846	Tizenhauz, 1846: pp. 452–453; Meyer, 1847: pp. 238–239; Decaisne, 1847: p. 313.
11. Jenischehir (Jenichehr)	January 1846	Tizenhauz, 1846: pp. 452–453; Teesdale, 1846: p. 230; Reissek, 1847: p. 195; Ludwig, 1866: p. 287 (no date); Murchison, 1864: p. 769.

Imperio Rutheno, published in 1739, included "Lichen coralloides, fruticosus, terrestris circa Oropolim [Orenburg or Chkalov] ad Iaicum [Jaïk or Ural] flumen..."[228] The explorer P. S. Pallas (1769) found "Lichen esculentus.... in aridissimus calcareis, gypseique montibus deserti tatarici, inter lapides crebra occurrit."[229] This formed the basis for several later descriptions.[230] F. Blume observed the lichen in the Caspian (Astrakhan') region ca. 1810–1812,[231] and likewise E. Eversmann (of the university of Kazan') in his journey (1820–1822) through the western Kirghiz steppe from Orenburg to Bukhāra.[232] Eversmann, in collaboration with L. Nees von

---

[228] Ammann, 1739: p. 176, no. 253.
[229] Pallas, 1771–1776: 3: p. 760; 1788–1793: 5: p. 516.
[230] Acharius, 1803: 2: p. 291 ("in montibus calcariis Sibirae"); 1810: p. 343 ("in montibus calcariis desert Tartarici"); Sprengel, 1825–1828: 4: p. 295 ("ad rubes calcarias in deserto tatarico").
[231] Noted in Treviranus, 1816: pp. 155–156; 1848: cols. 891–892; Eversmann, 1831: p. 355 n.; Elenkin, 1901b: p. 35.
[232] Eversmann, 1823: pp. 24–25 (27th October, 1820), 29, 116. See also Basiner (1842), 1848: pp. 65–66.

Esenbeck, published the first scientific account.[233] Fr. Goebel made the first chemical analysis, based on specimens collected by Friedrich Parrot in the Ararat region of northern Persia (1828).[234] To the east there are reports or collections from Sīstan,[235] Tien Shan,[236] Kashgaria,[237] and the steppes north of Lake Balkhash (near Karkaralinsk).[238] Similar information is available for north west Persia, Kurdistan and northern Syria;[239] southern Russia, including the Don valley,[240] Armenia,[241] Georgia,[242] Āzarbāījān,[243] the Caucasus[244] and the Crimea;[245] the Bulghaz Dagh and Lycaeonia (southern Asia Minor);[246] the region of Constantinople[247] and southern Greece.[248]

North Africa (Map 8)

Lecanora [Parmelia] esculenta has been reported from "North Africa"[249] and from "Algeria" or the "Algerian Sahara."[250] The early Arabs mention Barca (Cyrenaica)[251] and the Zāb of Qairwan (Tunisia), from which quantities were exported. Modern scientific observations commence with the French occupation of Algeria, and more particularly with the mission

---

[233] Eversmann, 1831: pp. 353, L. fruticulosa ("a montibus Mugosaricis [Mugodzhary] .... Emba fluvium .... deserta arenosa"), 355, L. affinis ("locum Araliensem septentrionem"), 356, L. esculenta ("deserti Tartarici .... desertis Kirgisicis").

[234] Goebel, 1830: pp. 393–399. Arnold ("Flechten auf dem Ararat," 1897) does not mention L. esculenta.

[235] Schlimmer, 1874: p. 12.

[236] Osten-Sacken and Ruprecht, 1869: p. 11; Elenkin, 1901b: pp. 32, 35–36.

[237] Elenkin, 1901b: p. 36.

[238] Meyer, 1830: pp. 359, 376. The distribution may extend into Mongolia and beyond (see Klement, 1965: p. 109).

[239] Boissier and Buhse, 1860: p. 242; Reichardt, 1864: p. 554; Schlimmer, 1874: p. 13 (Senjān, near Mosul); Steiner, 1921: p. 44 (Lecanora esculenta [Pall.] Evers. "Kalksteine der Steppe bei El Hammam unter Meskene am mittleren Euphrat, 300 m.").

[240] Pitra, 1868: pp. 7–8.

[241] Steiner, 1899: p. 293 (north of Erevan).

[242] Elenkin, 1901b: p. 32 (near Tiflis).

[243] Elenkin, 1901b: p. 32 (near Baku).

[244] Elenkin, 1901b: pp. 32, 35.

[245] Léveillé, 1842: pp. 139–140 (Sevastopol' - Balaklava); Walpers, 1851: pp. 317–318; Elenkin, 1901b: p. 36 (Theodossia [Feodosiya]); Tomin, 1956: p. 384.

[246] Krempelhuber, 1867: p. 600; Tschihatcheff, 1853–1869: 3, 2: p. 662.

[247] Riegler, 1852: 1: p. 110 ("Umgegend von Constantinopel"); Tschihatchcheff, 1853–1869: 3, 2: p. 662 (Byzantino).

[248] Marilaun, 1896: pp. 35–37 ("Guiona in Aetolien" [Aetolia]); Steiner, 1898: p. 107; Elenkin, 1901b: p.32 (Kiona [Kióni]).

[249] Treviranus, 1848: col. 893; Link, 1849: col. 731; Seeman, 1864: p. 205; Kolb, 1892: pp. 3–4; Shantz and Marbut, 1923: p. 83; Hooper, 1931: p. 325; Maurizio, 1932: p. 99; Hooper and Field, 1937: p. 135; Klement, 1965: p. 109.

[250] Anon., 1847: p. 816; Decaisne, 1847: p. 313; Hampe, 1848: col. 889; Munby, 1850: p. 71; Nylander, 1854: p. 312; J. Müller, 1858: pp. 89–90; Tschihatchcheff, 1853–1869: 3, 2: p. 662; Ludwig, 1866: p. 288; Flückiger, 1883: p. 28; Hue, 1891: p. 74; Marilaun, 1894–1895: 2: p. 810; Teesdale, 1897: p. 230; Sturtevant, 1972: p. 283; Meyerhof, 1947: p. 35; Bodenheimer, 1947: p. 2 (Atlas mountains). Eversmann, who spent the year 1859 in Algeria and visited Laghouat, apparently regarded the Asiatic and North African species as identical (Renard and Lacour, 1880: pp. 11–12).

[251] Cf. Trotter, 1915: p. 59 (kam-el-ouôta, "blé à la terre"); Hale, 1967: pp. 158–159. Pampanini (1914: pp. 320–321) only refers to Lecanora crassa Ach.

48

scientifique of 1840–1842. The results were published between 1846 and 1850 and include an important *flora* (J. F. C. Montagne). Here "Dr Lebrun" (? 1835) is said to have discovered *P. esculenta* in the Djebel A'mour region, west of Laghouat. Later observations extend the distribution to the north and the south, as far as the *wadi* Tamanrasset on the western margins of the Ahaggar massif. J. L. C. Guyon (1852) referred particularly to the highlands between Djebel Dira and Djebel A'mour, and to Beni-M'Zāb. The latter location is also mentioned in the botanical report (1857) of an expedition from Laghouat to Ouargla.[252] According to Guyon, the lichen was known as "excrément de la terre."[253]

(c) Falls of "manna lichen"

Aerial falls of "manna" have been reported from a number of places in western and west-central Asia, the majority in a zone extending from the southern Caspian to central Asia Minor, with a notable concentration around the headwaters of the Tigris and the Euphrates (Map 7). The reports are usually brief or otherwise unsatisfactory, but all appear to involve one or other of the species of *Lecanora*. Falls are sometimes explicitly associated with high winds and torrential rain, notably in the first quarter of the year. However, accounts of such sudden accumulations may be confused by the fact that the lichen is said only to "appear" to "grow" or to become readily observable after rain or heavy dew.

Lecanora spp. are insecurely attached to the soil or to rock, particularly at the close of the dry season. Fragments accumulate locally through the action of wind or running water and then, on comparatively rare occasions, may be carried for considerable distances through the atmosphere, finally to be deposited during rain storms or as the wind abates. Consequently the lichen is sometimes discovered in areas where it is otherwise unknown or at least unrecorded. In the circumstances the source areas cannot be exactly determined, but generally appear to lie to the east of the falls. During winter, atmospheric circulation is broadly controlled by a high-pressure system centred over Mongolia. Storms frequently develop around the spring equinox.

The earliest known reports of lichen "rain" relate to northern Persia or to neighbouring parts of Armenia in 1824, 1828, and 1829. In 1828 Professor Thénard brought to the notice of the French Academy of Sciences samples of deposits (forwarded by the French consul in Persia) from the Mount Ararat region (Map 7 [4]).[254] In places the substance covered the ground to a depth of about 15 centimetres. The phenomenon was not considered unique and Thénard briefly mentioned a similar fall in 1824.[255] F. Parrot, early in 1828, found comparable deposits at several localities in the same

---

[252] Cosson, 1857: p. 473. Similarly, Boutrekfine and Tilrhemt (shown in Map 8).

[253] Guyon, 1852: p. 212; O'Rorke, 1860: p. 417 (*ousseh elard*); Renard and Lacour, 1880: p. 15 (*oussokh el ard, oussak el ard*).

[254] Thénard, 1828: col. 55. Schlimmer (1874: p. 13) refers to Thénard as Ehénard.

[255] Noted by Göppert, 1831: p. 569; Reissek, 1847: p. 196; Ludwig, 1866: p. 287; Achundow in Muwaffiq ibn 'Alī, 1968: p. 357.

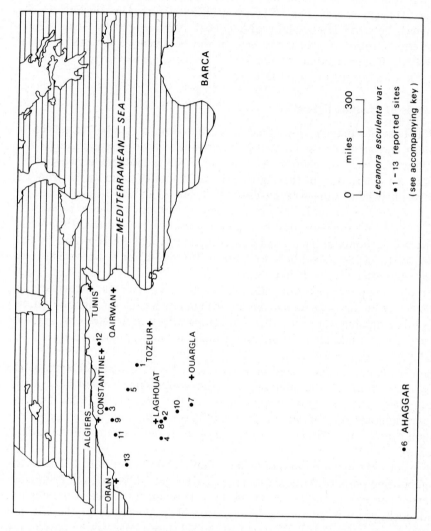

Map 8. *Lecanora esculenta* Evers., North Africa.

Reports of *Lecanora esculenta* Evers. in North Africa (Map 8). Numbers have no significance other than identification.

1. Biskra	Renard and Lacour, 1880 p. 3; Flagey, 1891–1892: p. 86.
2. Bou Trekfine	Cosson, 1857: p. 473.
3. Djebel Dira	Guyon, 1852: p. 212 [1835]; Ludwig, 1866: p. 288.
4. Djebel A'mour	Guyon, 1852: p. 212 [1835]; Montagne, 1846: pp. 294–295; Ludwig, 1866: p. 288.
5. Bou Saada	Renard and Lacour, 1880: p. 3; Faurel *et al.*, 1953: p. 314.
6. Tamanrasset	Faurel *et al.*, 1953: p. 314.
7. Beni-M'Zāb, Chebka du M'Zāb	Guyon, 1852: p. 212 [1835]; Cosson, 1857: p. 473; Reichardt, 1864: p. 555; Ludwig, 1866: p. 288; Faurel *et al.*, 1953: p. 314.
8. Laghouat	Kremelhuber, 1867: p. 603; Henneguy, 1883: p. 105.
9. Boghar (Titteri)	Walpers, 1851: pp. 317–318; O'Rorke, 1860: p. 417 [1849].
10. Tilrhemt	Cosson, 1857: p. 473.
11. Sersou (plateau)	Montagne, 1846: pp. 294–295.
12. Azeba	Flagey, 1896–1897: p. 52 (cf. Nylander, 1881: p. 183).
13. Oran (region)	Nylander, 1857: pp. 329–330; Flagey, 1896–1897 p. 52.

region.[256] In 1829, according to P. M. R. Aucher-Éloy, about 10 centimetres fell at a time of high winds in or around the city of Urmia (Reẓā'iyeh), south west of the Caspian (Map 7 [1]).[257] Here, however, "the people affirm [ed] that they had never seen this lichen before nor after that time" (Auchen-Éloy entered the region in 1830), and the fall was regarded as a miracle.

Several notices date from the 1840's. The *Journal de Constantinople* of January 26, 1846, mentioned falls in the spring of 1841 near Lake Van (Map 7 [3]) and at Sywrihissar (Sivrihisar) in central Asia Minor (Map 7 [12]).[258] These references to earlier events were prompted by reports of falls in the country around Jenischehir (Jenichehr) (Map 7 [11]) over a period of several days in January, 1846.[259] Then in March, during a violent storm, "manna" – evidently *Lecanora* sp. – settled on land near the town of Smorgonie (Smorgon' in Belorussiya) (Map 7 [10]).[260] Three years later (and possibly earlier) showers of lichen were reported near Byazid (Caldiran) (Map 7 [9]), north east of Van. They were investigated by a resident physician, Dr Heinig.[261]

---

[256] Goebel, 1830: pp. 393–399. Several later authors mention Parrot's discovery, but I can find no reference in his *Reise zum Ararat*, 1834.

[257] Aucher-Éloy, 1843: 2: p. 399 n. 1 (in "la collection d'Aucher-Éloy"; first noted by Léveillé, 1842: p. 141).

[258] Tizenhauz, 1846: pp. 452–453; Miquel, 1846: p. 416; Reissek, 1847: p. 195; Martius, 1852: cols. 20–21.

[259] Henneguy (1883: p. 105) gives 1845. Cf. O'Rorke, 1860: p. 417 (1845 at "Jenis-Bechir in the Crimea": ? an error).

[260] Tizenhauz, 1846: pp. 453–454. Noted by Meyer, 1847: p. 238. Decaisne (1847: p. 313) confuses Jenischehir with Smorgonie in "the government of Wilna." See also Gorski (1846–1847), 1912: pp. 76–77.

[261] Communicated by [Berkeley], 1849–1864: 1849: p. 581.

The first occurred at night on or about the 18th to the 20th April, "when there had been, for a whole fortnight, very rainy weather, with strong winds from the south-east and east-south-east." The tracts affected measured "from 5 to 10 miles each in circumference." Another deposit was "discovered" in June of the same year.

The next, largely unsubstantiated, references are from "Persia" and "near Erzurum" (Map 7 [7]) ca. 1854.[262] More widely reported was a fall, accompanied by a "gust of rain," in the vicinity of the village of Schehid Duzi, near Karput (Charput, Harput) (Map 7 [2]) in March, 1864.[263] The final known occurrence, around Diyarbakir (Map 7 [8]) in May 1890 was described in some detail by L. Errera.[264] Samples were obtained from the French consul in Aleppo. Again, the deposit was associated with a violent storm, and was apparently unknown locally.

(d) Lichen bread

Several accounts of "manna rain" concern areas in the grip of famine;[265] in less urgent circumstances, a fall might never have been reported. Where the substance was reasonably plentiful, a man could collect about .75 of a kilogram in an hour[266] or 4 to 6 kilograms in a working day.[267] Bread was made from the lichen alone or in combination with the meal of cultivated cereals. The food value is variable but generally small, for as much as 66 per cent consists of calcium oxalate.[268] The chief nutritional ingredient is lichenin or lichen jelly.[269] E. Lacour found a small amount of sugar (4 per cent) in a sample from Algeria,[270] and others have observed that the lichen has a slightly sweet taste.[271] It is known as *küdret-boghdasi* ("wonder grain") in

[262] Berkeley, 1857: p. 383; Moldenke, 1952: p. 127.

[263] Haidinger, 1864: pp. 129–130; 1865: pp. 170–177; Niessl, 1865: pp. 74–75. Murchison (1864: p. 769) got his information from Haidinger and he, in turn, obtained samples of the lichen from Baron Prokesch-Osten, the Austrian Internuncio at Constantinople. Cf. Seeman, 1864: p. 205; Reichardt, 1864: pp. 553–554; Ludwig, 1866; pp. 286–287; 1870: p. 52; Visiani, 1864–1865: pp. 284–306; 1867: pp. 197–205. 225–230. Murchison (*loc. cit.*) also mentions Malatia (Malatya) (Map 7 [5] and Diyarbakir (Diarbékir) (Map 7 [8]).

[264] Errera, 1893: pp. 83–91. Achundow in Muwaffiq ibn 'Alī, 1968: p. 357 (1891, without reference to place). Anon. (1891a: p. 255) reported falls at Diyarbakir and Merdin (Mardin, about 100 kilometres south east of Diyarbakir) in August, 1890.

[265] Léveillé, 1842: p. 140, and Aucher-Éloy, 1843: 2: p. 399 n. 1 (Urmia, 1829); Tizenhauz, 1846: p. 453 (Jenischehir, 1846); Berkeley, 1857: p. 383 (Erzurum, before 1857). Cf. Candolle, 1835: 2: p. 237 ("Lors de la disette de 1816 et 1817, on faisait dans les environs de Genève du pain de lichen.").

[266] [Berkeley] 1849–1864: 1849: p. 58 p. 581 (Byazid 1849, quoting Dr Heinig).

[267] Marilaun, 1894–1895: 2: p. 811.

[268] Goebel, 1830: pp. 393–399 (65. 91 per cent, northern Persia); Errera, 1893: p. 86 (57. 93 per cent, Diyarbakir); Lacour, 1880: pp. 449–453 (47 per cent inorganic, Algeria); Flückiger, 1883: p. 28 (22. 8 per cent calcium oxalate, 20 per cent other minerals); Tobler, 1925: p. 55.

[269] Errera, 1893: p. 89 (5 per cent); Lacour, 1880: p. 452 (10. 75 per cent); Tobler, 1925: p. 116. Cf. Goebel, 1830: p. 399 (23 per cent "Gallerte").

[270] Lacour, 1880: p. 452 (sucre incristallisable 2. 87 per cent, sucre cristallisable 1. 20 per cent); Renard and Lacour, 1880: p. 20.

[271] Munby, 1850: p. 71; Schlimmer, 1874: p. 12; Bodenheimer, 1947: p. 2 ("halva").

eastern Turkey,[272] and as *shīr-zād* ("milk-producer") in parts of Persia, where it is (or was) recommended as a gelactagogue for nursing mothers.[273] The Kurds called it "bread from heaven" or "bread from the earth,"[274] the Kazakhs of western Tartary "earth bread" (*semljanoi chleb*).[275]

The use of *L. esculenta* as food by man and beast (chiefly sheep) was probably most common among the pastoral nomads of the Kirghiz steppe[276] and of the high deserts to the south. Peripatetic groups were best able to take advantage of scattered and irregular accumulations of the lichen. The nomads of the desert of Sīstan were sometimes made aware of supplies by observing the feeding habits of antelope.[277] In the deserts of North Africa, on the other hand, the lichen only appears to have been regarded as an emergency food.[278]

In Armenia and Asia Minor, too, the lichen was consumed chiefly, if not exclusively, in times of scarcity and, from the evidence available, after heavy falls of the substance. Thénard (1828) provides the earliest evidence.[279] In 1829, according to Aucher-Éloy, the inhabitants of the region around Urmia collected the lichen to make bread, "qu'ils trouvérent assez bon et très-nourrisant."[280] This was an unusual expedient at a time of war and famine, and was prompted, apparently, by the way in which sheep devoured the lichen. Meal was prepared from the "manna" that fell around Van in 1841[281] and near Jenischehir in 1846.[282] At Byazid (1849) the lichen was "ground up with wheat and made into bread, or eaten simply in its raw

---

[272] Murchison, 1864: p. 769; Visiani, 1864–1865: p. 287; 1867: p. 199; Haidinger, 1865: p. 170; Ludwig, 1870: p. 52; Zopf, 1896: p. 187. Herbelot ([1697] 1777–1779: 2: p. 546) observed that "the manna of the Hebrews" was known as *kodret halvasi* (Turkish) and *haluat al kodrat* (Arabic).

[273] Schlimmer, 1874: p. 12 (*Chir-zadĕ*, more fully *Chirĕ ziadĕ Konèndèh*); Hooper, 1931: p. 325, and Hooper and Field, 1937: p. 135 (*Shīr-zād*); Meyerhof, 1947: p. 35. Pallas (1771–1776: 1: p. 366) mentions unspecified medicinal use in an area near the lower Jaïk (Ural) river.

[274] Dragendorff, 1898: p. 50 ("Himmelsbrod der Kurden"); Meyerhof, 1947: p. 35.

[275] Pallas, 1771–1776: 1: pp. 366, 382; 1788–1793: 1: pp. 571, 595. According to O'Rorke (1860: p. 417), the lichen is known as *takaout* in "Arabia," but I have found no reference to its presence in the peninsula, apart from northern Syria and Palestine (Hue, 1891: p. 74). Berber *tâkaoût* = "galle de tamarix" [*T. articulata*] (Duveyrier, 1864: pp. 172–174; Salmon, 1906: pp. 10, 48). See also Leclerc in Ibn al-Baiṭār, 1877–1883: 1: p. 302; 2: p. 405; 3: p. 25; Basset, 1899: p. 59 (*takout* [in Ibn al-Baiṭār] "c'est le nom berbere donne a l'euphorbe dans le Maghreb central."); Leclerc in 'Abd ar-Razzāq (18th century), 1874: pp. 83, 347 (Berber *tâkoût* = *Tamarix* sp. [galls] and *Euphorbia* sp.)

[276] Pallas, 1771–1776: 1: p. 366; Eversmann, 1823: pp. 24–25.

[277] Schlimmer, 1874: p. 12.

[278] Montagne, 1846: pp. 294–295 (in years of famine "Les Ouled-Naïl on fait avec le lichen et l'orge un pain grossier, mais assez substantiel.") See also Smith, 1921: p. 405 (camels, gazelles and other quadrupeds); Hale, 1967: pp. 158–159 (sheep, in Libya). Renard and Lacour (1880: p. 15) quote reports (1854, 1879) that the lichen was locally regarded as poisonous and that shepherds were quick to sell animals that had eaten the substance. The explanation seems to be that, taken in quantity, it is highly indigestible.

[279] Thénard, 1828: col. 55. Willemet (1787) and Amoreux (1787) do not refer to *L. esculenta* among the "lichens économiques."

[280] Aucher-Éloy, 1843: 2: p. 399 n. 1.

[281] Reissek, 1847: p. 195.

[282] Tizenhauz, 1846: p. 453. See also Visiani, 1867: pp. 197–205, 225–230 (Karput, 1864). According to Decaisne (1847: p. 314), a "hail of manna" served as food for the inhabitants of Herāt during a siege.

state."[283] Similarly at Diyarbakir (1890) lichens were mixed with flour in the ratio of 3 : 1.[284]

Knowledge of lichen bread was put to advantage during the French campaign in Algeria.[285] A field surgeon, Dr Raymond (ca. 1845) recognised *L. esculenta* as "the lichen of the steppes of Tartary" and drew this to the attention of the army commander, General Jusuf. The latter prepared a report on the lichen (May 11th., 1847) which is said to have appeared annually – apparently *in situ* – during or just after the rainy season. Horses were fed for a period of three weeks on a mixture of barley and lichen "without ill effect." Subsequently, bread was baked using the lichen alone and also mixed with flour.

L. esculenta must have been collected and consumed on numerous, unrecorded occasions. Among farming folk it has been essentially a famine food with low nutritional value. For the pastoral peoples of west-central Asia, on the other hand, it may have been a fairly regular, if minor, item of diet. Medicinal or therapeutic use has only been reported in modern times from Sīstan and western Tartary.

F. QUERCUS spp.

Manna has been collected from several species of oak. The most frequently reported are *Quercus persica* Jaub. et Spach and *Q. brantii* Lindl.[286] The combined distribution extends from Turkish Armenia in the north and west, south eastward through the Kurdo-Zagrosian mountains to Farsistan (Map 9).[287] *Q. mannifera* Lindl. (*Q. sessiliflora* Sm. var. *mannifera*) also belongs to Armenia and Kurdistan.[288] Other manniferous species include *Q. vallonea* Kotschy (*Q. aegilops* L. ssp. *vallonea*), *Q. tauricola* Kotschy, *Q. graeca* Kotschy, and *Q. calliprinos* Webb (*Q. coccifera* L. var. *calliprinos*).[289]

---

[283] [Berkeley] 1849–1864: 1849: p. 581, quoting Dr Heinig.
[284] Errera, 1893: p. 84.
[285] Montagne, 1846: pp. 294–295. See also Guyon, 1852: p. 212.
[286] Djavanchir-Khoie (1967: p. 137) regarded *Q. persica* and *Q. brantii* as separate species. According to Zohary (1973: 2: p. 358), "[*Q. persica*, *Q. oophora* Kotschy and *Q. vesca* Kotschy] can be looked upon as varieties of the polymorphic *Q. brantii* [Lindl.]" (thus *Q. brantii* ssp. *persica*, or *Q. brantii* ssp. *brantii* var. *persica*; cf. O. Schwarz, 1936: p. 19; Zohary, 1961: p. 174; Rechinger, 1971: p. 14).
[287] Based on Lindley (Brant), 1840: p. 40: Jaubert and Spach, 1842–1857: 1: Tab. 55; Kotschy, 1862: Tab. XVIII, XXI; Boissier, 1867–1888: 4: p. 1173; Nábělek, 1929: p. 23; Guest, 1933: p. 81; O. Schwarz, 1936: p. 19; Camus, 1934–1948: 1: pp. 526–528; Gauba, 1949–1953: 59: p. 52; Zohary, 1950: p. 42; 1961: p. 174; 1973: 2: pp. 358, 582; Sabeti, 1966: no. 622; Djavanchir-Khoie, 1967: pp. 133–137; Rechinger, 1971: p. 14).
[288] Lindley (Brant), 1840: pp. 40–41; Tschihatchcheff, 1853–1869: 3, 2: p. 465; Boissier, 1867–1888: 4: pp. 1164–1165 (*Q. sessiliflora* var. *mannifera*); Masters in Lindley and Moore, 1870–1874: 2: p. 951; Nábělek, 1929: p. 22; Wehmer, 1929: pp. 139–140; Hooper, 1931: p. 335; O. Schwarz, 1936: p. 12; Hooper and Field, 1937: p. 162; Takhtadzhiana, 1954–1966: 4: p. 385. In Persia it may be confused with *Q. persica* (Sabeti, 1966: no. 639; Djavanchir-Khoie, 1967: p. 193).
[289] *Q. vallonea* ("chêne du Kurdistan", *drakht gheizeilefi*, Morgan, 1894–1905: 1: p. 32; *gez ʻelfī*, Wulff, 1966: p. 76) is often mentioned along with *Q. persica*. For *Q. tauricola*, see Hooper, 1931: p. 335; Hooper and Field, 1937: p. 16; for *Q. graeca*, Wehmer, 1929: p. 137; and for *Q. calliprinos*, Camus, 1934–1948: 1: p. 463. Cf. Ocampo, 1900–1901: pp. 407–420 (Memoría sobre el *Quercus mellifera*).

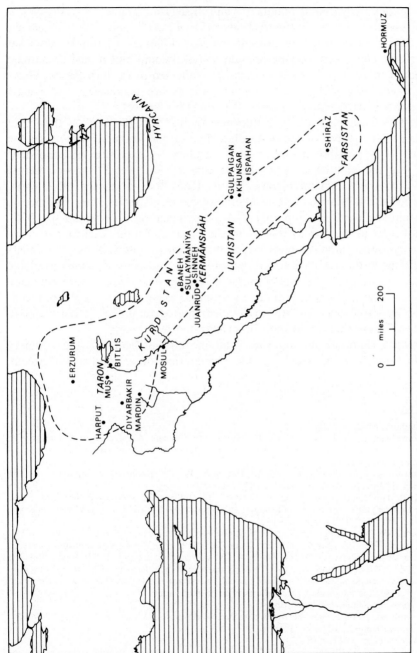

Map 9. Combined distribution of *Quercus persica* Jaub. et Spach, *Q. brantii* Lindl., and *Q. mannifera* Lindl.

Oak manna is probably available in many parts of south west Asia, but it is chiefly associated with eastern Anatolia and Kurdistan, whence the descriptions "Diyarbakir manna"[290] and "Kurdish manna."[291] In Persia it is known as *gaz* or *gezú*[292] (Kurdish *ghezo* or *gezza*)[293] and more particularly *gaz-ālāfi*.[294] *Gaz-angubīn*, literally "tamarisk honey [manna]," also has generic meaning and may be applied to oak manna and to the sweetmeat prepared therefrom.[295] Both mannas are collected in western Persia. *Quercus persica* is doubly useful, for the acorns are comestible and widely appreciated, particularly at times of poor grain harvest.[296]

Gaz-ālāfi accumulates to a thickness of half a centimetre on the upper surface of the leaves of the oak rather than on the branches. Like tamarisk manna, it is possibly a purely insect product, an excretion or honeydew. This was the opinion of several 19th-century observers,[297] most recently supported by F. S. Bodenheimer.[298] A. Haussknecht, on the other hand, regarded the substance as an exudation of the host species after puncture by an aphid (? *Coccus manniparus*).[299] Others, including S. Moghadam (1930), do not refer to the presence of an insect, but rather imply that meteorological conditions (notably high day temperatures, followed by cool misty conditions in the early morning) determine the quantity of manna available which, in any event, is highly variable.[300] The dry "tears" were collected between June and August, either by simply shaking or beating the branches of the host tree over a cloth, or by emersing cut branches in hot water to give a solution that could then be evaporated to a syrup.

Purified (filtered) manna was used in folk medicine,[301] the ordinary product much more widely in the preparation of a popular sweetmeat.

---

[290] Hanbury, 1876: p. 287.

[291] Bodenheimer, 1947: p. 6. Curiously enough, the ancient country or nation known as *Manna* (*Mana, Manai, Mannai*) lay in Kurdistan, south of Lake Urmia [Reẓā 'īyeh] (Herzfeld [*Itinerary of Šarrukēnu*], 1968: p. 234).

[292] Rawlinson, 1839: p. 104; Layard, 1887: 1: pp. 349–350 (Bakhtiari of Luristan).

[293] Rich, 1836: pp. 142–143; Wright, 1847a: p. 350.

[294] Moghadam, 1930: p. 113 ("manne d'herbe." '*alaf*). Chemical analyses in Berthelot, 1863: p. 85; Ludwig, 1870: pp. 33–35; Flückiger, 1872: pp. 159–164; 1883: p. 26; Ebert. 1908: pp. 480–481; Wehmer, 1929: pp. 139–140.

[295] Zénob de Glag (Klag), 1867: p. 355 (*kazabîn = guezengubin, guezenbo*); Olivier, 1801–1807: 2: pp. 359–360 (*guiésén-guébin*); Dupré, 1819: 1: pp. 93–94 (*guez-inguèbin*); Ferrier, 1856: p. 26 (*guzengébine*); Tozer, 1881: p. 263 (*ghezenghi*); Layard, 1887: 1: pp. 349–350 (*gazenjubin*); Hooper and Field, 1937: p. 162 (*gazenjubeen*). Hooper (1909: p. 33) observed that the term *gazangabin* was "loosely applied"; elsewhere (1931: p. 334) he gives the name *pune* (Tehrān) for the "green cake" prepared from oak manna.

[296] Djavanchir-Khoie, 1967: p. 136.

[297] Rawlinson, 1839: p. 104; Ferrier, 1856: p. 26 (edit. note); Layard, 1887: 1: pp. 349–350; possibly also Kinneir, 1813: p. 329.

[298] Bodenheimer, 1947: p. 6.

[299] Haussknecht, 1870: pp. 244–245 (*manna quercina*; the name *tréhala* is incorrectly applied); also maintained by Djavanchir-Khoie, 1967: p. 136.

[300] Brant (H.M. consul in Erzurum) quoted by Lindley, 1840: p. 40; Brant (1838), 1841: p. 352; Berthelot, 1863: p. 85; Bodenheimer, 1947: p. 6.

[301] Brant quoted by Lindley, 1840: p. 40 (gelactagogue); Schlimmer, 1874: p. 358; Collin, 1890: p. 104; Moghadam, 1930: p. 117; Hooper, 1931: pp. 334–335.

William Ouseley (1819) referred to this as *dúsháb* (Persian).[302] Other descriptions include *kudret helvasi* (Turkish)[303] and *mann as-samā* (Arabic),[304] "divine/heavenly sweetmeat/manna." However, these names, like *gaz-angubīn*, may also refer to similar preparations containing other kinds of manna.

There are allusions to oak manna in Assyrian and in Classical sources from the time of Theophrastus (*supra* pp. 5, 9). References to the province of Hyrcania all appear to be based on accounts of Alexander's expedition. A. F. von Stahl, with close personal knowledge of the region, commented on the statement by Curtius Rufus, thus: "on the northern side of the ridge Alexander entered the plain of Bandar-i-Gaz (Shore of Gaz), a small town and port on the Caspian Sea. *Gaz* or *Giaz* is the Persian term for a sort of manna; a sweet juice which, as a result of perforation by insects, flows out of the leaves of a certain kind of oak, and when dried is like sugar. This *gaz* is largely collected in Kurdistan, and at Isfahan is used for the manufacture of a sort of sweet called *gaz*."[305]

Zénob de Glag's *History of Daron* (Taron or Taraunitis, to the west of Lake Van) contains an interesting observation. Bishop Zénob (died 323/4) lived in the monastery of Sourp Garabed near the border of Syria and Armenia. Taron, he wrote, "produit en grande abondance des pâturages et du miel. Semblable à la manne qui tombaît du ciel pour les Juifs, une rosée plus douce que le miel se dépose ici sur les arbres; on l'appelle *kazabîn*."[306] This appears to be the earliest notice of oak manna from a region mentioned by several later authorities. The 12th-century Armenian physician Heratsi Mekhithar gives *sbidag kazben*, "white manna."[307] Among the Arabs, Ibn Butlān (died 1052) referred to Diyarbakir[308] and Ibn al-Baiṭār (1197–1248) to the manna that "falls from the *djîdâr*" (*Quercus ? coccifera*),[309] presumably in Armenia or Kurdistan. Al-Muqaddasī (ca. 985) mentioned that

---

[302] Ouseley, 1819–1825: 1: p. 453. Cf. Haussknecht. 1870: p. 245 (*pekmes*, ? Kurdish); Hooper, 1909: p. 33 (*dustab*); Moghadam, 1930: p. 118 (*baslogh*).

[303] Fahir İz and Hony, 1952. Cf. Rich, 1836: 1: pp. 142–143 (*kudret halvassi*); Brant quoted by Lindley, 1840: p. 40 (*koodret-helvahsée, ghiok-helvahsée*, "sweetmeat of heaven"); Ainsworth (1837), 1868: p. 501 (*kudrat halvassi*); Haussknecht, 1870: p. 244, and Flückiger, 1883: p. 26 (*küdret halwa*); Balfour, 1885: 2: pp. 852–853 (*kudrat-ul-halwassi*); Dragendorff, 1898: p. 166 (*kudrat halwa*).

[304] Guest and Townsend, 1966: p. 497.

[305] Stahl, 1924: p. 326. Dorostkar's monograph (1974: p. 118) on the forests of eastern Hyrcania [Gurgan] lists only three species of oak, none of them, so far as is known, manniferous: *Q. castaneifolia* C. A. Meyer, *Q. macranthera* Fischer et Meyer, and *Q. iberica* Steven. Thomas Herbert travelled in Persia 1627–1628 and reported of Hyrcania (1928: p. 169): "that tree called *occhus*, which is said [by Pliny] to distil honey we found not; but one that had a sweet sap, or juice, which 'tis likely gave the occasion of that report. ..." The only potentially manniferous species reported from eastern Hyrcania is *Salix caprea* (*supra*, p. 60 n. 329).

305 Zénob de Glag (Klag), 1867: p. 355. The history was written in Syriac and later translated into Armenian.

[307] Mekhithar, 1908: p. 163, no. 158.

[308] Ibn Butlān, 1531: p. 24 ("Manna vero cadit super arbores landri, et glandium, in regione Sagiuar et Dyarbether.")

[309] Ibn al-Baiṭār, 1877–1883: 1: p. 390, no. 546; Sontheimer, 1840–1842: 1: pp. 275–276 (*dschidār*).

obtainable from Mosul (al-Mauṣil).[310] In the land of *Huz* – probably *Hazar*, north east of Mosul – Fr. Odoric of Pordenone (1286–1331) "found manna of better quality and in greater abundance than in any part of the world."[311]

Oak manna was known to European botanists and students of *materia medica* from at least the beginning of the 17th century. Johann Bauhin (died 1613) included *quercum mel* among a variety of mannas;[312] his brother Gaspard (died 1624) stated that it was available in the Appenines.[313] Pierre Pomet associated oak manna with the hinterland of Hormuz, whence it was exported, packed in goat skins, to Goa.[314] This seems to be based on Garcia da Orta, a resident of Goa and author of *Colloquies on the simples and drugs of India* (1563).[315] Sir John Chardin (ca. 1666–1673) may refer to oak manna in his *Voyages en Perse*,[316] and likewise Niccolao Manucci who travelled in India (1653–1708) and records that Shāh 'Abbās (1642–1667) presented to the emperor Aurangzīb "a sealed box of gold, full of manna from the mountains of Shīrāz."[317]

Among 18th-century travellers, Karsten Niebuhr (1761–1764) is the most informative.[318] He observed "[La manne] s'attache à Merdîn (Mardin) comme une farine sur les feuilles de certains arbres qu'on appelle *ballôt* et *afs*[319] .... et que je crois être des chênes. .... Tous s'accordoient à assurer, qu' entre Merdîn et Diarbekr on la recueilloit principalement sur les arbres qui produisent la noix de galle, c. à. d. sur des chênes. La récolte de cette

---

[310] Al-Muqaddasī, 1901: p. 235, no. 1001. Chau Ju-kua, in his work (*Chu-fan-chi*) on Chinese and Arab trade in the 12th and 13th centuries, wrote (Hirth and Rockhill, 1911: p. 140) under Wu-ssï-li (Mosul): "In autumn there falls a heavy dew which, under the action of the sun's rays, hardens into a substance like powdered sugar. This is gathered and is sweet, pleasant tasting food with purifying and cooling properties; it is real *kan-lu* ("sweet dew," that is manna). The editors identify this as the manna of *Alhagi camelorum* or *A. maurorum* (*tar-angubīn*), but oak manna is also possible. On *kan-lu* from Sou-lin (Sūristān, Syria/Iraq), see Chavannes and Pelliot, 1911–1913: 1: p. 131.
[311] Odoric, 1891: p. 59; 1913: p. 109. Cf. Mendeville (ca. 1350), 1953: 1: p. 109. This may be the region to which Leonhard Rauwolf ([1573–1576], 1693: p. 205) referred ("another sort of manna .... from Armenia").
[312] J. Bauhin, 1650–1651: 1: p. 180.
[313] G. Bauhin, 1671: p. 497 (under "manna officinarum"); followed by Johnstone, 1662: p. 337. The only other known reference to oak manna in Europe (Provence) is in Garidel, 1719: p. 391 ("le miel qu'on trouve sur les feuilles du chêne ...."). Cf. Geoffroy, 1741: 2: p. 590; Sestini, 1788: p. 92.
[314] Pomet, 1694: 1: p. 239.
[315] Orta, 1913: p. 281 ("Another kind [of manna] is sometimes seen in Goa, liquid in leather bottles, which is like coagulated white honey. They sent this to me from Ormuz"). Followed by Acosta, 1578: p. 309; Clusius, 1605: p. 164. Manna was on sale in Surāt in the middle of the 17th century (Thevenot [1633–1667], 1949: p. 25). Garcia da Orta also mentions "another kind [of manna] in large pieces mixed with leaves, coming by way of Baçora [Basra]." Cf. Acosta and Clusius *locs. cits.*; Linschoten (1583–1592), 1885: 2: pp. 100–101; Teixeira (1586–1605), 1902: p. 204; Stevens, 1715: p. 30. Balfour (1885: 2: pp. 852–853) remarked upon manna "carried to the market in Mosul in goatskins."
[316] Chardin, 1811: 3: pp. 295–296 ("la troisième sorte de manne"). The first kind seems to be that of *Alhagi maurorum*; the second Chardin identifies as from the tamarisk. Cf. the comments of Frederick, 1819: pp. 252–253.
[317] Manucci, 1907–1908: 2: p. 51.
[318] Niebuhr, 1773: pp. 128–129.
[319] According to Löw (1967: 1: p. 626), *ballūṭ*, '*afṣ* = Q. aegilops L., "vallonia oak." See also Guest and Townsend, 1966–1974: 3: p. 497 (probably *Q. infectoria* Oliv. [*Q. lusitanica* Lam.]; possibly also *Q. aegilops*). Virey (1818: p. 125) refers to *Q. ballota* Desf. (*Q. ilex* L.).

manne se fait à Merdîn en Août, ou suivant d'autres en Juillet, et on la dit plus abondante après un certain brouillard fort épais, ou pendant un tems humide, que pendant les jours séreins. On ne soigne pas ces arbres aux environs de Merdîn, mais lorsque la manne tombe, en cueille qui veut dans le bois, sans en demander, ni acheter la permission du gouvernement." Jean Otter (1748) briefly described the collection, before dawn, of manna from "trees" in the vicinity of Gulpaigan, 50 kilometres north of Khunsar.[320] G. A. Olivier (ca. 1793), too, made local enquiries concerning *guiésen-guébin*, found "en abondance" in Kurdistan and northern Persia on "un arbre de moyenne grandeur, ou un grand arbrisseau ressemblant un peu au chêne,"[321] presumably a reference to the dwarf oak *Q. persica*.

There are a dozen or more first-hand accounts of the collection, preparation and marketing of oak manna in Luristan and, more particularly, in the Kurdo-Armenian cultural province, from the beginning of the 19th century to almost the present day. As late as 1947, according to F. S. Bodenheimer, "thousands of kilograms" were collected in Kurdistan alone.[322] The product was consumed fresh in rural areas, as a substitute for honey and sugar, and was also stored "in large quantities" for winter use.[323] Peasants in the vicinity of Diyarbakir are said to have used it "instead of butter in cooking."[324] Through trade manna was available in urban centres between Erzurum and Shīrāz. Several travellers refer to the towns of Muş, Bitlis and Karput (Elâzíǧ) in Turkish Armenia,[325] and to the "export" of oak manna, "compressed into a solid mass," from various parts of Kurdistan[326] and Luristan, where the nomadic Bakhtiyaris appear to have been particularly involved.[327] Unfortunately, in unscientific descriptions of the commercial product there is some risk of confusion with the alimentary manna found (in the same region) on *Tamarix gallica*. The two were probably used together in the preparation of *gaz-angubīn* and similar sweetmeats.

G. SALIX spp.

Bīd-angubīn (*angubīn*, "honey") or *bīd-khecht*[328] is a manna exuded, under unusually hot and dry conditions, from the leaves and young branches

---

[320] Otter, 1748: 1: p. 197. Presumably the oak rather than tamarisk.
[321] Olivier, 1801–1807: 2: pp. 359–360.
[322] Bodenheimer, 1947: p. 6.
[323] Wright, 1847a: p. 351.
[324] Hanbury, 1863: p. 108; 1876: p. 287.
[325] Kinneir, 1813: p. 329; Burckhardt, 1822: p. 601; Brant (1838), 1841: p. 352; Koch (1843–1844), 1846–1847: 2: p. 407; Tozer (ca. 1879), 1881: p. 263, 269, 303; Bishop, 1891: p. 351; Lynch, 1901: 2: p. 151. See also Porter (1817–1820), 1821–1822: 2: p. 471.
[326] Rabino, 1911: pp. 13 (Senna or Sanandaj), 24 (Baneh), 46 (Juvanrūd or Juanrūd); *gazangebin* here = oak manna.
[327] Dupré (1807–1809), 1819: 1: pp. 93–94 (Mosul); Rawlinson (1836), 1839: p. 104; Ferrier, 1856: p. 26 (Kermānshāh province); Layard, 1887: 1: pp. 349–350; Rabino, 1916: p. 6; P. Schwarz, 1896–1936: 7: p. 883 n. 8. For Kurdistan/Luristan generally, see Frederick (1813), 1819: p. 257; Rich, 1836: 1: pp. 142–143; Wellsted, 1838: p. 48; Ainsworth (1837), 1868: p. 501 (Sulaimānīyah); Polak, 1865: 1: p. 286; Dragendorff, 1898: p. 166.
[328] For *khecht*, see *supra* p. 37.

of species of willow (Persian *bīd, bed*), notably *Salix fragilis* L. (1753).[329] The latter is found between western Europe and central Asia[330] and southward to a zone (wherein the manna has chiefly been reported) extending from southern Europe, Asia Minor,[331] Palestine and Syria,[332] through Armenia and Persia,[333] to western Tibet and the north-western provinces of India.[334] *S. fragilis* is sometimes cultivated[335] and has been widely valued for the real and assumed medicinal properties of the bark, sap, leaves and seeds. The pulverized bark is still employed in the Near East as a febrifage and must have been of greater importance before quinine (prepared from the bark of *Cinchona* sp.) became available from the middle of the 17th century. Interestingly, Ignatius Molina in his *History of Chile* (first published in 1787) observed that "the country people .... make use of the bark [of *S. chilensis*, or *S. humboldtiana* Willd.] which they believe possesses a highly febrifugal quality."[336] The same species also "yields annually a great quantity of manna." This is one of the comparatively few references to the product from the New World, but whether the bark and the manna were used before the Spanish conquest has not been determined.

Both Pliny and Dioscorides (1st century A.D.) mention the medicinal properties of the concrete "juice" or "gum" of the *salix*,[337] and so too do medieval Persian and Arab writers, including Muwaffiq ibn 'Alī (ca. 970),[338] Ibn Sīna (980–1037),[339] and Ibn al-Baiṭār (1197–1248).[340] They may refer to what later was described as "manna." There is a more direct allusion in the *Mirabilia Descripta* of Friar Jordanus (ca. 1300–1330). He observed of the region around Tabrīz that "on a kind of willow are found certain little worms which emit a liquid which congeals upon the leaves of the tree, and also drops upon the ground, white like wax; and that excretion is sweeter than honey and the honeycomb."[341] Joannes Bodaeus à Stapel (ca. 1630), a

---

[329] Also, apparently, *S. caprea* L. (Dymock, 1890–1893: 3: pp. 364–365); *S. babylonica* L. (Bretschneider, 1882–1895: 3: no. 328); *S. chilensis* [*S. humboldtiana* Willd.] (Molina [1787], 1809: 1: p. 137); *S. tetrasperma* Roxb. and *S. persica* Boiss. (Moghadam, 1930: p. 70). *S. fragilis* = Persian *bīd-bīd-khechti* (Morgan, 1894–1905: 1: p. 32), *bīd-ḥeštī* (Wulff, 1966: p. 75).

[330] Gueldenstadt (1768–1773), 1787–1791: 1: pp. 113, 196; Sprengel, 1825–1828: 1: p. 99; Levshin, 1840: p. 88; Ledebour, 1842–1853: 2: p. 598; Steven, 1856–1857: 29, 2: p. 389; J. D. Hooker, 1872–1897: 5: p. 630.

[331] Tschihatchcheff, 1853–1869: 3, 2: p. 485; Boissier, 1867–1888: 4: p. 1185.

[332] Boissier, 1867–1888: 4: p. 1185; Dinsmore and Dalman, 1911: p. 202; Post, 1932–1933: 2: p. 530.

[333] Tschihatchcheff, 1853–1869: 3, 2: p. 485; Boissier, 1867–1888: 4: p. 1185; Morgan, 1894–1905: 1: p. 32; Gilliat-Smith and Turril, 1930: 10: p. 485; Sabeti, 1966: no. 744; Takhtadzhiana, 1954–1966: p. 353.

[334] J. D. Hooker, 1872–1897: 5: p. 630; Burkill, 1909: 71; Bamber, 1916: p. 20.

[335] According to Irwin (1839–1840: 8, 2: p. 892), "[manna] is produced in Toorkistan on the dark barked or cultivated willow ...."

[336] Molina, 1809: 1: p. 137.

[337] Pliny, 1961–1968: 8: pp. 45–47; Dioscorides, 1952–1959: 3: p. 89.

[338] Muwaffiq ibn 'Alī, 1968: p. 195, no. 175, under *chillâf*.

[339] Ibn Sīna, 1608: 1: p. 402.

[340] Ibn al-Baiṭār, 1877–1883: 3: p. 10 ("le suc de la feuille ....", under *ghareb*); see also *ibid*: 2: p. 43 (*khilâf* and *safsâf*).

[341] Jordanus, 1863: p. 8. The Dominican Fr. was bishop of Columbum (Malabar).

Dutch botanist and commentator of Theophrastus' *Historia Plantarum*, also took the view that insects (*culices, crassi*) deposited manna on willows.[342] It is said that insects, including bees, collect the saccharine substance,[343] which may have led to a misunderstanding.

Willow manna has been reported from regions as far apart as southern France, northern India,[344] and China ("gum" of *S. babylonica* L., used in medicine).[345] The annual harvest is very variable.[346] The only known description of the mode of collection (in August and September) is provided by S. Moghadam in his *Mannes de Perse* (1930). The branches of the tree are scraped with the blade of a knife and the drops (*gouttelettes*) of manna are allowed to fall into a sack of "flour." The latter prevents the substance forming a glutinous mass. Rose petals may be used to impart a fragrance and, incidentally, give the red appearance that was observed by A. Ebert.[347]

Gaspard Bauhin (1560–1624) included the product of the willow (from the Appenines) under "manna officinarum."[348] A quantity was collected in the summer of 1754, which was notably hot and dry, in the vicinity of Carcassonne.[349] The region most commonly mentioned in the literature of the 19th and 20th centuries is northern Persia (Map 10), more particularly the Elburz range and the province of Tehrān (Damāvand, Shariar).[350] W. Ainsley (*Materia Medica*, quoting a Persian source) refers to *beed khusht*, "a variety of manna found on a willow in Khorāsān."[351] In William Ouseley's *Travels* (1819–1825) we find that "about Hamadān [manna] settles on the *bīd* or willow."[352] Jordanus's early reference to the region of Tabrīz has already been mentioned. Further west again, "a syrupy fluid, which in taste and appearance exactly resembled oak manna" was observed ca. 1840 on willows in the neighbourhood of Lake Van.[353] Edward Frederick's account of *gez* or "Persian manna" (1813) suggests that that found upon the *beod* was particularly esteemed.[354] At all times *bid-angubīn* seems to have been

---

[342] Bodaeus à Stapel, 1644: p. 151. Cf. Lindley, 1840: p. 40.
[343] Cadet, 1810: p. 130 (France); Polak, 1865: 2: p. 287 (Elburz region, Persia). Dymock (1885: 1: p. 62) thought that the manna obtained from tamarisk and oak, as well as the willow, was the result of insect punctures.
[344] Dey, 1896: p. 133.
[345] Bretschneider, 1882–1895: 3: no. 328. *S. fragilis* was used in China in the preparation of charcoal for gunpowder (Chun, n.d.: p. 48).
[346] Moghadam, 1930: pp. 69–70.
[347] Ebert, 1908: p. 479.
[348] G. Bauhin, 1671: p. 497; followed by Johnstone, 1662: p. 327.
[349] Duhamel du Monceau, 1758: 1: p. 152 (valley of the Fresquet, near Pennautier); Marcorelle, 1760: p. 501. See also Cadet, 1810: p. 130 (a young tree, exposed to the sun – in France, but location undisclosed).
[350] Polak, 1865: 2: p. 287; Kiepert (Haussknecht), 1868: p. 473; Ludwig, 1870: p. 44; Schlimmer, 1874: p. 359 (Chehriar); Raby, 1889: p. 205 (vicinity of Tehrān, and Hérouz-Kouh [? Harhāz Kuh] to the north east); Collin, 1890: p. 104 (Chehriar); Ebert, 1908: pp. 479–481; Moghadam, 1930: p. 69 (Shariar, district in the province of Tehrān, to the south west of the capital). A. H. Wright (1847a: p. 351) of the American mission at Urmia, also mentions the manna of "a species of willow growing on the water courses in Persia."
[351] Ainsley, 1826: 1: p. 210; apparently followed by Royle, 1839: 1: p. 345 (*bed-khisht*).
[352] Ouseley, 1819–1825: 1: p. 453.
[353] James Brant (H.M. consul in Erzurum) in Lindley, 1840: p. 40.
[354] Frederick, 1819: p. 257.

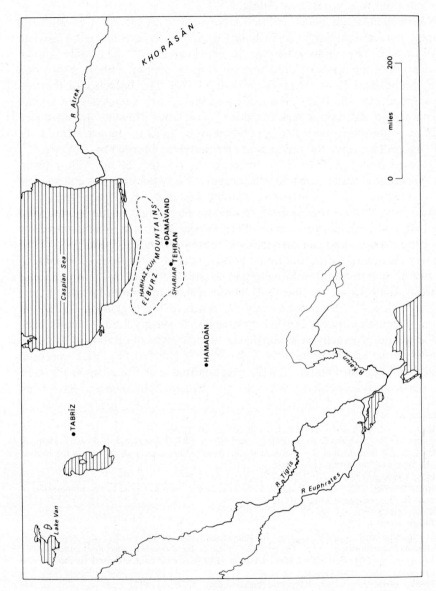

Map 10. Principal area of collection of willow manna, *bīd-angubīn* or *bīd-khecht.*

used largely, if not exclusively, in folk medicine.[355] Sold in native pharmacies, it probably reached a market far beyond the chief areas of collection.

H. TAMARIX spp.

(a) Uses of Tamarix spp.

Members of the genus *Tamarix* have been employed in a variety of ways from very remote times. In Pharaonic Egypt the shrub had both religious and medicinal importance.[356] The *Papyrus Ebers* (compiled ca. 1550 B.C. from earlier sources) refers to the leaves and the juice in prescriptions and lists of drugs.[357] In ancient Mesopotamia the alkaline ashes were used in "washing rituals."[358] Arab and Persian physicians of medieval and later times prescribed the galls or "fruit" (Arabic *tamr-el-aṭl*, Persian *gazmāzak*, *gazmāzū*, *kāzmazāk*).[359] In Algeria the dried and powdered leaves and roots, as well as the fresh leaves and young shoots, are (or were until recently) ingredients in folk medicine.[360] The wood, leaves and sap of *T. chinensis* Lour. (*ch'eng liü*) are found in Chinese *materia medica*.[361]

In North Africa and western and central Asia the galls of *Tamarix* spp. have long been used in tanning (likewise the bark) and in dyeing.[362] The wood has served for making implements and a variety of craft products.[363] Rafters of *aṭl* (*T. aphylla* [*articulata*]) were remarked by Charles Doughty.[364] Among the Tartars, branches of *T. gallica* (known as *gilghine*) were particularly valued for the handles of whips.[365] Everywhere tamarisks

---

[355] Chiefly a pectoral. See Collin, 1890: p. 104; Moghadam, 1930: p. 71. Chemical analyses in Ludwig, 1870: pp. 45–46; Raby, 1889: pp. 205–208; Ebert, 1908: pp. 479–481.

[356] Loret, 1887: p. 36, no. 88.

[357] Ebbell, 1937: pp. 81, 85, 97, 106, 131. Cf. Dioscorides (*Herbal*), 1934: pp. 61–62.

[358] R. C. Thompson, 1949: pp. 39 ff.; Levey, 1959: p. 123. Tamarisks are generally tolerant of salt, growing by the sea-shore and in saline depressions, and the ashes contain much sulphate of soda.

[359] Ibn Sīna (908–1057), 1608: 1: p. 402; Angelus, 1681: p. 256 (*qours ghezmazegi*, "fructus tamaricis"); Watt, 1889–1893: 6, 3: pp. 410–412 (India); Meyerhof, 1918: p. 203; Kirtikar and Basu, 1918: 1: pp. 138–139; Ducros, 1930: p. 32, no. 56; Meyerhof, 1931: p. 53; Al-Ghāfiqi (ca. 1160), 1932–1938; 1: pp. 69–70; Maimonides (1135–1204), 1940: pp. 9 (ed. comm.), 100, no. 200; Ibn Kaysān (died 1582), 1953: p. 28; Al-Samarqandī (died 1222), 1967: p. 71 (*kharmāzaj*); Muwaffiq ibn 'Alī (ca. 970), 1968: p. 180; Al-Ṭabarī (9th century), 1969: p. 290, no. 470.

[360] Hilton-Simpson, 1922: pp. 65, 71, 72. See also Duveyrier, 1864: p. 174.

[361] Regnault, 1902: p. 152, no. 107; Laufer, 1919: p. 348; Read and Liu Ju-Ch'iang 1927: no. 239; Stuart, 1928: p. 428; Hooper, 1929: p. 133. On *T. chinensis*, see Bunge, 1835: p. 102; Bretschneider, 1882–1895: 2: pp. 364–365; Forbes and Hemsley, 1886–1888: 23: p. 347.

[362] Bellew, 1864: Appendix X; Brandis, 1874: p. 23; Watt, 1889–1893: 6, 3: pp. 410–412; Dey, 1896: p. 311; Dastur, 1962: pp. 46–47, 152–153.

[363] Pottinger, 1816: p. 62; Burckhardt, 1829: 2: p. 215 (*ithel*); J. D. Hooker, 1854: 1: p. 392; Bellew, 1864: Appendix X; Brandis, 1874: pp. 21, 23; Aitchison, 1891: p. 203; Boucheman, 1934: pp. 38, 61, 82, 92; Doughty, 1936: 2: p. 20; Dastur, 1962: pp. 46–47, 153.

[364] Doughty, 1936: 1: pp. 185, 638; 2: p. 560.

[365] Radcliffe, 1789: p. 108. Cf. Aitchison, 1891: p. 204 (western Afghanistan and north east Persia).

are cut for fuel, a fact perhaps first reported by Abū al-Fidā (1273–1331) from the desert regions of Bukhāra where two forms of the shrub (*ghaḍâ* and *ṭarfâ*) were found.[366] It is also widely planted as an ornamental,[367] the larger species as shade trees in the hot lands of North Africa and western Asia.

Such variety of use over a long period of time indicates close observation of the tamarisk and tends to support the general view that the manna, although available only locally, was also recognised and exploited at an early date.

(b) Species of Tamarix

Folk nomenclature

The Persian word *gaz* or *gazm* refers to the genus *Tamarix*, whence *gaz-angubīn*, "tamarisk honey" or manna.[368] The same name or some compound (for example, *gaz-khera*, *siahgaz*) is found in Baluchistān,[369] Sindh[370] and Afghanistan.[371] In the memoirs of Bābur, founder of the Mogul dynasty, there is a reference (1493–1494) to the Dara-i-Gaz, "the valley of tamarisk" near Balkh in northern Afghanistan.[372] In Pushtu we find *pirghaz* and *ghwaz* (*T. articulata*),[373] in Panjabi *ghazlei* (*T. gallica*),[374] and in Kurdish *gazo* or *gezu*.[375] In Turkī the generic name is *yulgun*.[376]

In Arabic there are two principal names, which to some extent are interchangeable. *Ṭarfā* (Spanish *taraje*) more usually refers to *T. gallica* (*mannifera*),[377] *aṭl* (Assyrian *ashlu*, Hebrew *ashel*, *eshel*) or some cognate form to *T. aphyla* (*articulata*, *orientalis*).[378]

---

[366] Abū al-Fidā, 1848–1883: 2, 2: p. 212. See also Brandis, 1874: pp. 21–23; Aitchison, 1891: p. 203; Bretschneider, 1898: p. 986; Tate, 1909: p. 88; Gamble, 1915–1936: 1: p. 67; Dastur, 1962: pp. 46–47.

[367] Belon du Mons, 1555: p. 104 (the gardens of Cairo), and many later travellers.

[368] *Gez* or *gaz* was also a Persian unit of length (Le Strange, 1890: p. 49; Temple in Mundy, 1905–1936: 2: p. 67 n.). Maunsell (1896: p. 239) observed that *gez* was the name of "a species of wild silk .... found in the hills near Jezire (Kurdistan)."

[369] Burkill, 1909: pp. 12–13.

[370] Brandis, 1874: p. 21.

[371] Aitchison, 1886–1887: p. 467; 1891: pp. 87, 203. In the southern region of Helmand, tamarisk is (also) known as *kirī*; according to Gamble (1915–1936: 1: p. 67), Tamil *kiri* = *T. gallica*.

[372] Bābur (1483–1530), 1912–1921: 1: p. 14. An earlier edition (1826: p. 7, n. 3) translates "the valley of Gez or Manna."

[373] Bellew, 1864: p. 238, and Appendix X (*T. orientalis* = *T. articulata*).

[374] Brandis, 1874: p. 20.

[375] Laufer, 1919: p. 348 n. 7. According to Rich (1836: 1: pp. 142–143), *ghezo* is the name of the manna.

[376] Laufer, 1919: p. 348, n. 7. Cf. Anon., 1929: p. 18 (*yūlghūn*); Bedevian, 1936: no. 3348 (Turkish *manna ilǧun aǧ*). For the many vernacular names of peninsular India, see Brandis, 1874: pp. 20–22; Gamble, 1915–1936: 1: p. 67; Bamber, 1916: p. 112. Sanskrit *jhāvuka*, tamarisk.

[377] Muwaffiq ibn ʿAlī (ca. 970), 1968: p. 231. Cf. Ibn Sarābī (? 12th century), 1541: p. 46; Forskål, 1775a: p. LXIV; Dinsmore and Dalman, 1911: p. 21; Löw, 1967: 3: pp. 402–403 (*T. nilotica* var. *mannifera*). Forskål also gives *hattab achmar*, followed by Muschler, 1912: 1: p. 648; Post, 1932–1933: 1: p. 224 (*hatab ahmar*); Bedevian, 1936: no. 3348 (*hhatab ahhmar*).

[378] Forskål, 1775a: p. LXIV; ʿAbd ar-Razzāq (18th century), 1874: p. 21; Huber, 1891 (technical index); Meyerhof, 1918: p. 203; Guest, 1932: p. 17; Meyerhof in Maimonides, 1940: p. 9; Lewin in al-Dīnawarī, 1953: p. 20; Löw, 1967: 3: p. 398.

Scientific nomenclature and distribution

The genus *Tamarix*, comprising approximately 90 species, extends from the Atlantic islands to central Asia (Saharo-Arabian and Irano-Turanian geobotanical provinces).[379] Reports of deposits of manna generally fail to identify the species satisfactorily. *T. gallica* L. (*T. troupii*) is most frequently mentioned, but this is a cosmopolitan species, or rather group of species, that occurs almost throughout the range of the genus.[380] From the known evidence, it has not been possible to map the distribution of *T. gallica* var. *mannifera* Ehrenberg (1827) (Fig. 8), also known as *T. mannifera* Ehr. ex Bunge (1852) and apparently including *T. nilotica* var. *mannifera* Schweinf.[381] K. H. Rechinger gives *T. mannifera* var. *persica*.[382] These species or varieties may be broadly associated with the central areas of tamarisk-manna production, from the Sinai peninsula to western and central Persia (Map 11).

Manna has also been reported on *T. senegalensis* D. C. (?*T. gallica* var. *senegalensis*), known as *mboundou* in Senegal itself. A. Sébire (1899) observed: "Aurait-il une écorce astringente et fébrifuge et donnerait-il une manne détergente et expectorante comme le *T. gallica* var. *mannifera*....."[383] *T. gallica* (*T. canariensis* Willd.) is also present in the Atlantic islands,[384] and from Palma we have Roger Barlow's interesting reference (ca. 1550) to "a certain dewe clammy like honey which the people do gather in bagges of lether and after putteth it in erthen pottes.... Thei call it mangula [Spanish *mangla*, "gum"] ...."[385] This was probably tamarisk manna.

Another group of manniferous species is found in the extreme southeastern part of the range – Sindh, western Punjab, Baluchistān,[386] southern Afghanistan and eastern Persia (Sīstan). The group consists of:

T. dioica Roxb. According to D. Brandis (1874)[387] and G. Watt (1893)[388], *T. dioica* produces manna (*maki*) in Sindh.

T. pentandra Pall., 1788 (*T. pallasii* Desv., 1825). This is known as *shingir gaz* and *shōra-gaz* in Baluchistān[389] and as *gez-e-māzej*, "the manna

---

[379] Zohary, 1973: 2: pp. 385–386. Bunge (1852) described 51 species, Niedenzu (1895) 65 species.

[380] The *Index Kewensis* (4: 1895, and supplements) places under *T. gallica–T. pentandra* Pall., *T. pallasii* Desv., *T. mannifera* Kotschy, *T. nilotica* Ehr. ex Bunge, *T. indica* Willd., *T. senegalensis* D.C., *T. canariensis* Willd., *T. anglica* Webb. Several of these species/varieties have been described as manniferous.

[381] Löw, 1967: 3: pp. 402–403. *T. mannifera* is illustrated in Haynald, 1894: tábla 2.

[382] Rechinger, 1964: p. 13.

[383] Sébire, 1899: p. 25.

[384] Webb, 1841: pp. 428–429; Bunge, 1852: p. 62 (Tenerife); Oliver *et al.*, 1868–1937: 1: p. 15.

[385] Barlow, 1932: p. 102.

[386] References to the tamarisk manna of India (Dey, 1896: p. 311; Kirtikar and Basu, 1918: 1: pp. 38–39; Chopra, 1933: p. 532) presumably relate to the North West Frontier.

[387] Brandis, 1874: p. 23.

[388] Watt, 1889–1893: 6, 3: p. 411 (collected and used for confection).

[389] Hooper, 1909: pp. 34–36; 1931: p. 340; Hooper and Field, 1937: p. 176. Aitchison (1891: p. 97) identified *gaz-shōra* as *T. tetragyna* Szovits ex Bunge (*T. octandra* Bunge) of western Afghanistan and north east Persia.

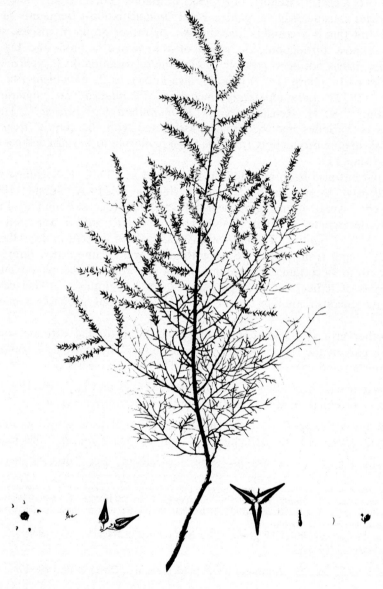

Fig. 8. *Tamarix (gallica) mannifera* Ehren. (Ehrenberg and Hemprich [1820–1825], 1930: Taf. II).

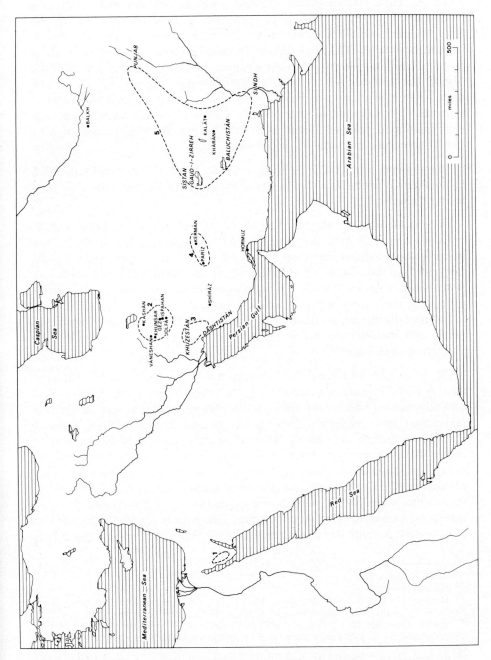

Map. 11. Tamarisk manna reported, 1–4 on *Tamarix gallica* (*mannifera*) Ehren., 5 on *T. dioica* Roxb., *T. pentandra* Pall., and *T. aphylla* (*articulata*) Karst.

67

tamarisk," in Persia.[390] "Late in the spring the *shōra-gaz* of the Gaud-i-Zireh [a saline (*shōra*) marsh in north west Baluchistān and south west Afghanistan] yields in very large quantities a kind of sugar which the Baluchis call *tirmi*. This occurs on the branches in round lumps as large as walnuts or smaller. The flock owners (*maldar*) collect a quantity of it, and a large number of men from Bandar i-Kamal Khan and Rūdbār also visit the district for the purpose .... The branches of the tamarisk are cut off and dried. When dry they are beaten with wooden mallets until the sugar is separated."[391] Deposits of "earth sugar" (*shira-i-zamin*), found in periodically inundated parts of Sīstan, appear to be derived from manna shed by tamarisk bushes, probably *T. pentandra*.

T. macrocarpa Ehr. (1827) ex Bunge (1852), known as *kiri* and *gaz-surkh* in northern Baluchistān. The "saline accretion" on the branches may be a form of manna.[392]

T. articulata Vahl, 1781 (Fig. 9) (*Thuja aphylla* L., 1755, 1759, and *T. orientalis* Forsk., 1755), now known as *T. aphylla* (L.) Karst. This is the *khora-gaz* of Afghanistan and the *kiri*, *siahgaz* and *shakar-gaz*, "sugar tamarisk," of Baluchistān.[393] Between Kalāt and Péshtar Khan in north east Baluchistān, Charles Masson (1826–1838) observed "the variety of tamarisk producing the saccharine gum called *shakar-gaz*."[394] D. Brandis (1874) found that during hot weather this species also yielded manna (*misri lei*) in Sindh and the Punjab, where the product was used medicinally and to adulterate cane sugar.[395]

The Tuareg of the Ahaggar collect manna from *T. aphylla* (*tabarekkat*), as well as from *T. gallica* (*azaoua*).[396] According to M. Gast (1968), "[les] concrétions sont très recherchées par les enfants qui les sucent au fur et à mesure de leur cueillette; elles servent parfois à sucrer le thé. Mise à bouillir, la manne de tamaris devient un peu amère. Sa consommation suscite de nombreux rots. On s'en sert aussi de sirop pour arroser la galette brisée en morceaux dans un plat, ou simplement comme boisson rafraîchissante. Cette manne ne peut se conserver; une fois récoltée, elle se transforme assez vite, devient brune et gluante. .... En Ahaggar .... la manne est considérée comme un fortifiant bon pour la croissance des enfants et favorisant la santé des personnes adultes." This appears to be the manna

---

[390] Wulff, 1966: p. 76.
[391] Hooper, 1909: p. 36.
[392] Hooper, 1909: p. 34; S. G. Harrison, 1951: p. 411. Cf. *T. salina* Dyer, the *ghwa* of the North West Frontier, said to yield "manna [used] as a mild aperient" (Bamber, 1916: p. 112).
[393] Burkill, 1909: pp. 12–13; Hooper, 1909: p. 34; 1931: p. 340; Hooper and Field, 1937: p. 176. Aitchison (1891: p. 87) gives *gaz-shakar* (*T. gallica* var.) from Afghanistan and north east Persia.
[394] C. Masson, 1842: 2: p. 116. From Khairan (Khārān) to the south west of Kalāt "*gaz-shakar* [again used in the sense of manna] .... is said to be collected from a variety of *T. gallica*" (Aitchison, 1891: p. 129), but the species may have been misidentified.
[395] Brandis, 1874: pp. 22–23.
[396] Nicolaisen, 1963: p. 177; Gast, 1968: pp. 242–243. For *T. articulata*, see Gubb (*La flore saharienne*), 1913: pp. 112–115, and Ozenda (*Flore de Sahara*), 1977: p. 347.

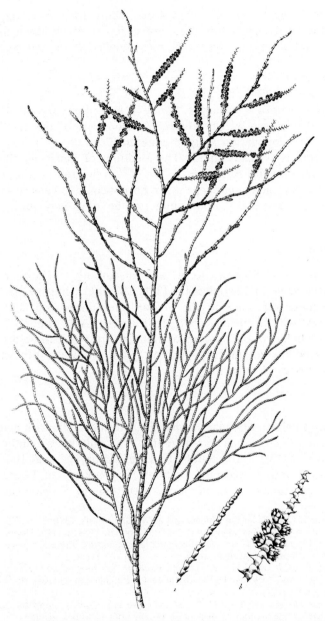

Fig. 9. *Tamarix articulata* Vahl (Vahl, 1790–1794: 2: tab. XXXII).

mentioned by Leo Africanus (1525) in his account of the Targa (Tuareg) people of the southern Ahaggar. In John Pory's translation (1600) we read: "Not farre from Agadez (Agades) there is found great store of manna, which the inhabitants gather in certain little vessels, carrying it while it is new unto the market of Agadez; and this manna, being mingled with water they esteeme very daintie and pretious drinke. They put it also into their pottage, and being so taken, it hath a marvellous force of refrigerating or cooling, which is the cause that there are so few diseases, albeit the ayre of Tombuto (Timbuktu) and Agadez be most unwholesome and corrupt."[397]

(c) Gaz-angubīn

Persian *gaz-angubīn* ("tamarisk honey") and *gaz-i-shakar* ("tamarisk sugar"), is also known as *gaz khūnsār*[398] from one of the chief areas of production. The appellation *khushk-angubīn* ("dry honey")[399] is confusing for the manna is typically moist. The Bedouin of Sinai call the product simply "manna of tamarisk," *menn-et-tarfā*.[400] In (western) Persia, the names *gaz-angubīn* and *gazambū* appear to have acquired generic status and other mannas, particularly that found on the oak, may thereby be included.[401]

Ancient times

The tamarisk was well known to the ancient authors.[402] There are particularly full comments in Theophrastus (370–287 B.C.), but nowhere is there an unambiguous reference to the manna. Herodotus (5th century B.C.) comes nearest with his *méli ek myríkes* in eastern Asia Minor.[403] Accounts by western scholars of the "manna of Sinai" (*infra* pp. 72ff) before the early modern period do not identify the host plant. However this was almost certainly a species of tamarisk.

The Middle Ages

The Arab and Persian pharmacists specifically refer to tamarisk manna more rarely than to alhagi manna (*tar-angubīn*). The only known authorities are al-Bīrūnī (ca. 1030),[404] Ibn Sarābī (? 12th century),[405] and Ibn al-Baiṭār (1197–1248)[406] if it is accepted that his "dry honey" of Persia was the product of the tamarisk. In addition, according to S. Moghadam, *gaz-*

---

[397] Leo Africanus (Al-Hassan ibn Muḥammad al-Wezāz al-Fāsi), 1896: 3: p. 799.

[398] Collin, 1890: p. 104; Moghadam, 1930: p. 103; Hooper, 1931: p. 340; Hooper and Field, 1937: pp. 175–176 (*T. gallica* var. *mannifera*, and *T. pentandra*). There is a place called Gez, 15 kilometres north-north-west of Ispahan.

[399] Ibn al-Baiṭār, 1877–1883: 2: p. 32 (*khochkendjubîn*). See Moghadam (1930: p. 85) on the descriptions "wet" and "dry" applied to the mannas of alhagi and tamarisk respectively.

[400] Meyerhof, 1947: p. 35.

[401] *Supra*, p. 59. Hooper and Field (1937: p. 175) give *gaz-ālāfi* as one of the names of the manna of *T. gallica* var. *mannifera*. It is more usually applied to oak manna.

[402] Strabo (ca. 63 B.C.–24 A.D.), 1960–1969: 7: p. 309 (in Arabia); Dioscorides (ca. 78 A.D.), 1934: pp. 61–62; 1952–1959: 3: pp. 71–72; Pliny (23–79 A.D.), 1961–1968: 4: pp. 166–167 (in Italy); Galenus (2nd century A.D.), 1821–1833: 12: pp. 80–81. See also Fraas, 1845: pp. 109–110; Lenz, 1859: pp. 640–641; Langkavel, 1866: p. 23, no. 87.

[403] *Supra*, p. 8. Other possible references to tamarisk manna are mentioned on p. 10.

[404] Al-Bīrūnī, 1973: 1: p. 310, quoting Ibn Maṣāh [Ibn Māsawaih, died 857] ("it [apparently the manna of the Bible] is *jaranjubīn* and is the dew that falls on the tamarisk tree.")

[405] Ibn Sarābī, 1531: p. 59 ("Manna cadit super tamarisci"). See also Guignes (Ibn Sarābī), 1905: 6: pp. 58–59.

[406] Ibn al-Baiṭār, 1877–1883: 2: p. 32. Cf. Laufer, 1919: pp. 347–348; Guest and Townsend, 1966–1974: 3: p. 498.

angubīn was included among the mannas of Persia by Zein el-Āttar who lived in the 14th century.[407]

After 1500

Over the period 1500 to 1800, statements concerning tamarisk manna were either very brief, and probably based on the earlier Arabic literature, or, when made by travellers, tended to be ambiguous. Altomarus (1562)[408] and Johann Bauhin (ca. 1600)[409] do little more than mention the product.[410] Garcia da Orta (1563) may have seen it (or oak manna) "in Goa, liquid in leather bottles, which is like coagulated white honey. They sent this to me from Ormuz [Hormuz], for it corrupts quickly in our land ...."[411] *Guesengebin* or *tamaricis mel* is included with several other oriental mannas in the *Pharmacopoea persica* (Paris, 1681) of Fr. Angelus.[412] John Chardin (1660's) associated the product with the ancient province of Susiana and more particularly with the coastal lands (Khuzistān and Dashtistān) at the head of the Gulf.[413] It is known that the tamarisks of Khuzistān occasionally yield manna.[414]

Tamarisk manna is found in widely separated regions. The Atlantic islands and Senegal, the Ahaggar plateau of the central Sahara, Khuzistān in south west Persia, and a zone stretching from Sīstan to the north-western margins of peninsular India have already been identified. The remaining regions lie in Persia and Arabia. There are apparently no conclusive reports of tamarisk manna from central Asia (north of Persia and Afghanistan)[415] or from China.[416] Apart from the presence of particular species and varieties of tamarisk (representing a small fraction of the genus), it has been suggested that the manna may be the result of localized climatic and/or edaphic conditions, but these have not been determined. More particularly, it is claimed (i) that the substance exudes from the branches of the tamarisk after punctures by insects (*Coccus manniparus*),[417] and (ii) that in Sinai at least it

---

[407] Moghadam, 1930: p. 12. Zein el-Āttar ("the pharmacist") was born in Shīrāz.

[408] Altomarus, 1562: p. 10.

[409] J. Bauhin, 1650–1651: 1: p. 181.

[410] See also Maffei (15th century), 1559: p. 627.

[411] Orta, 1913: p. 281 (not *tar-angubīn*, which is discussed separately). Followed by Acosta, 1578: p. 309. Moghadam (1930: pp. 13, 104) claims that Orta mentions tamarisk manna.

[412] Angelus, 1681: p. 359.

[413] Chardin, 1811: 3: p. 295 ("Il croît en abondance dans la province de Sousiane, et particulièrement autour de Daurac .... qui est l'Araca [Aracia] de Ptolémée.") Cf. the statements by Fryer (1672–1681), 1909–1915: 2: p. 201, and I. E. Fabri, 1776: pp. 136–139 (*Manna Chardinii Tamariscina*).

[414] Moghadam, 1930: p. 106.

[415] Hooper (1909: p. 34) mentions central Asia, as well as Arabia, Persia, Baluchistān and Afghanistan.

[416] According to Bretschneider (1882–1895: 2: no. 527), there are three species of tamarisks in China (*T. chinensis* Lour., *T. juniperina* Bunge, *T. pallasii* Desv.). See also Bunge, 1835: p. 102; 1852: pp. 45–46; Franchet, 1883–1888: 5: pp. 206–207; Forbes and Hemsley, 1886–1905: 23: p. 347. F. P. Smith (1871: p. 144) and Stuart (1928: p. 259) give what they claim to be the Chinese name for tamarisk manna (*ch'êng-ju*), but this is rejected by Laufer, 1919: p. 348.

[417] Moghadam, 1930: p. 103. Schlimmer (1874: p. 359) suggested that the coccus might be introduced more widely to increase the supply of manna.

is a wholly animal product, an insect excretion or honeydew (*infra*, pp. 78–79). There may be significant physical and chemical differences between the products of different regions.[418]

Persia: From the middle of the 18th century, if not earlier the territory around Khūnsār was celebrated for tamarisk manna,[419] which was used medicinally[420] and more especially in confectionary. It was collected as far to the north west as Kāshān[421] and south to Jolfā[422] and Ispahan which specialized in the making of a sweetmeat (also known as *gaz-angubīn*)[423] resembling nougat. During the hot season, three harvests might be taken from the same area at intervals of up to ten days.[424] The coated twigs were usually simply beaten over an outstretched cloth or an earthen vessel. Edward Frederick (ca. 1810), after a determined search, found manna in the vicinity of Khūnsār. He agreed with local opinion that it was an insect product, but also clearly stated that "the tamarisk bears no resemblance to the *gavan*, the bush on which the *gez* is found."[425] Either he was entirely mistaken or, perhaps, the *gavan* was a species of *Alhagi* or *Astragalus*.

References to *gaz-angubīn* in Kurdistan, without indication of the host plant,[426] should probably be interpreted as oak manna. Tamarisk manna was, however, produced in at least one other part of Persia, around Kermān and Pārīz. "The speciality of Pārīz is *gaz*," observed P. M. Sykes (1906);[427] and in the neighbourhood of Kermān, according to J. E. T. Aitchison (1891), it was "obtained in large quantities [from *T. gallica* var. *mannifera*] and exported in all directions."[428] India was the chief market outside Persia.[429]

Arabia Petraea: *Tamarix gallica* (*nilotica*) var. *mannifera* has been recorded for Palestine, western and southern Arabia (Jiddah and the

---

[418] Moghadam, 1930: pp. 103–104.
[419] Schlimmer (1874: pp. 358–359), followed by Collin (1890: p. 104), claimed that it was only produced here. See also Otter, 1748: 1: p. 197; Polak, 1865: 2: pp. 285–286; Haussknecht, 1870: p. 246; Moghadam, 1930: pp. 103–104 (between Khūnsār and Fāridān [? Vāneshan], "où le Tamarix pousse à l'état sauvage").
[420] Moghadam, 1930: p. 105.
[421] Arnold, 1877: 1: p. 295.
[422] Wills, 1883: p. 158. Bunge (1852: p. 63) reported *T. mannifera* Ehr. as far south as Shīrāz.
[423] Recipes also included flour, rose water, cane sugar or honey, almonds and pistachios. See Malcolm, 1815: 2: p. 562 n.; Binning, 1857: 1: p. 333; Schlimmer, 1874: 2: p. 358; Arnold, 1877: 1: p. 295; Wills, 1883: p. 158; Curzon, 1892: 2: p. 502; Sykes, 1906: p. 433; Moghadam, 1930: p. 105.
[424] Mounsey, 1872: pp. 190–191; Moghadam, 1930: p. 104.
[425] Frederick, 1819: p. 251. Cf. comments by Tabeeb, 1819: p. 268; Hardwick, 1822: pp. 182–186. John Malcolm (1815: 2: p. 562 n.) also claimed that "this [tamarisk] honey is produced by an insect." See also Mounsey, 1872: pp. 190–191.
[426] Oliver (ca. 1793), 1801–1807: 2: p. 360; Frederick, 1819: p. 253 (Kermānshāh); Ouseley, 1819–1825: 1: p. 452; Ferrier, 1856: p. 26. Moghadam (1930: p. 19 n. 2) maintained that tamarisk manna was not produced in Kurdistan. The product is mentioned, but without reference to where in Persia it was collected, by Fraser, 1834: p. 465; Binning, 1857: 1: p. 333; Flückiger and Hanbury, 1879: p. 415; Dymock, 1885: 1: p. 61; 1890–1893: 1: p. 159; Wehmer, 1929: p. 203.
[427] Sykes, 1906: p. 433.
[428] Aitchison, 1888–1894: p. 42; 1891: p. 204. See also Polak, 1865: 2: p. 286. Hooper (1909: p. 34; 1931: p. 340) and Hooper and Field (1937: p. 176) follow Aitchison.
[429] Hooper, 1909: p. 36; Watt, 1889–1893: 6, 3: p. 412; Markham in Orta, 1913: p. 281 n. 2 (from Persia and Arabia).

Ḥaḍramawt), Arabia Petraea (including Sinai) and neighbouring parts of Egypt.[430] However, reliable reports of tamarisk manna (*menn eṭ-ṭarfā*) are available only from a number of localities in south-western Sinai – the *wadis* Gharandal, Feirān, El Sheikh, Taib, Isla (Esle), Nasb, and possibly Solaf and Hebran (Map 12).

The "manna, produced by insects (*Coccus manniparus*), brought from Mount Tabor [northern Palestine]" and exhibited in London ca. 1846[431] has not been more fully described or explained. J. L. Burckhardt, when travelling through El Ghor (the valley of the Jordan), heard of "beirūk honey" that, in May and June, "dropped from the leaves and twigs of a tree called *gharrab*,"[432] possibly *Salix babylonica* L.[433] or *Populus euphratica* Oliv.,[434] but certainly not the tamarisk. In Sinai, Alfred Kaiser (1924) found saccharinè deposits on *Artemisia herba alba* Asso. and two species of *Haloxylon, articulatum* Bunge and *schweinfurthii* Aschers.[435] This manna, known respectively as *menn esh-shīḥ* and *menn er-rimt*, was collected by the local Bedouin, but in only very small quantities. *Alhagi maurorum*, although present in the region, is not known to yield manna.

Tamarisk manna may have been brought to ancient Egypt (? *mennu-t ḥet'*), along with frankincense, from the Arabian peninsula (*supra*, p. 6). The celebrated manna of the Wilderness of Sin[436] – between Elim (Wadi Gharandal) and Mount Sinai or Horeb – was first associated with an observable phenomenon by Flavius Josephus in his *Antiquitates Judaicae* (ca. A.D. 93–94). "And to this very day, all that region (Sinai) is watered by a rain like to that which then, as a favour to Moses, the Deity sent down for men's sustenance. The Hebrews call this food *manna*; for the word *man* is an interrogative in our language, asking the question 'what is this' ?"[437] Whether or not we have here the most apposite etymology (*supra*, p. 4), there is no reason to suppose that the Israelites, while in Egypt, were familiar with imported manna and, in any event, they could hardly have been aware of its mode of origin. Josephus's reference to "all that region" may be simply an exaggeration; but it is also probable that there have been adverse ecological changes over the last 2000 years and that stands of tamarisk along the *wadis* are now much less extensive than they were.[438]

---

[430] Decaisne, 1834–1835: 3: p. 260; Bunge, 1852: p. 63; Boissier, 1839–1845: 1: p. 775; Woenig, 1886: p. 341; Volkens, 1887: p. 107; Hart, 1885: p. 426; 1891: p. 84; Niedenzu, 1895: p. 9; Muschler, 1912: 1: p. 648; Blatter, 1919–1933: p. 73; Löw, 1967: 3: pp. 402–403; Post, 1932–1933: p. 224.
[431] Westwood, 1846: p. 659.
[432] Burckhardt, 1822: pp. 392–393 ("The Arabs eat [the manna] like honey, with butter; they also put it into their gruel, and use it in rubbing their water skins, in order to exclude the air.")
[433] Sprengel, 1807–1808: 1: pp. 270, 380 (*gharb, gharab*); Guignes (Ibn Sarābī), 1905: 5: p. 499 ("Le saule d'Egypte porte encore le nom de *gharab*"); Bedevian, 1936: no. 3035 (*gharb*). Cf. Ibn al-Baiṭār, 1887–1883: 3: p. 10, no. 1631 (*ghareb*).
[434] Guest and Townsend, 1966–1974: 3: p. 498.
[435] Kaiser, 1924: pp. 111 ff. Hart (1885: p. 434) reported *A. herba alba* from the Wadi El Sheikh among other localities.
[436] Exodus. 16; Numbers. 11; Deuteronomy. 8.
[437] Josephus, 1967–1969: 4 [1]: pp. 331–335. Contrary to the statement of Laborde (1841: p. 95), Josephus does not specify the *tamarisk* manna of Sinai.
[438] Ebers, 1872: p. 232; H. S. Palmer, 1906: pp. 216–220; S. G. Harrison, 1951: p. 409.

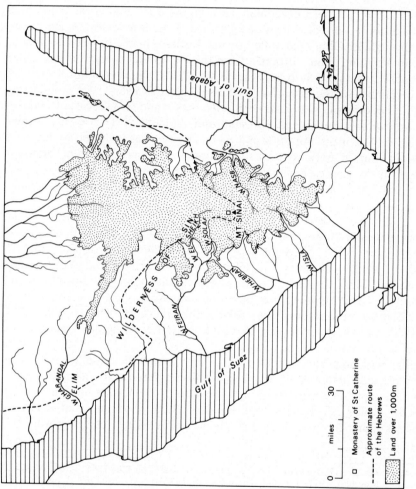

Map 12. Peninsula of Sinai.

Gulf of Aqaba

Gulf of Suez

MT SINAI   M.HASB

W.SOLAF

W.EL SHEIKH

WILDERNESS OF SIN

W.FEIRAN

W.HEBRAN

W.ISLA

W.ELIM

W.GHARANDAL

□   Monastery of St Catherine

----   Approximate route
        of the Hebrews

⠄⠄⠄   Land over 1,000m

0   30
|----|----|----|
      miles

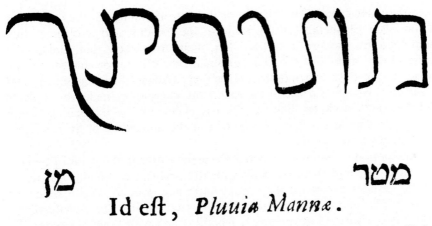

מטר   מן

Id eſt, *Pluuia Mannæ*.

Fig. 10. "Rain of Manna" Inscription, Sinai (P. Thomae Obecini Novariensis, in Kircher, 1652–1654: 2: p. 120).

Among early medieval writers, Antoninus Martyr [Placentinus] (ca. 570) provides perhaps the first circumstantial description of the manna of Sinai.[439] The Byzantine historian Georgius [Syncellus] (ca. 700) appears to have thought that it was much the same as the manna brought from Parthia.[440] In the course of the central and later Middle Ages a number of travellers visited southern Sinai and refer to the local manna, but without associating it with the tamarisk or indeed with any particular plant. Leonardo Frescobaldi (1384) and Fr. Francisco Suriano (1494) each brought back flasks of manna from the monastery of St. Catherine.[441] Neither, apparently, saw the substance *in situ*. Arnold von Harff (1497) reported that it "[fell] each year in the high mountains round about, some six miles from the monastery, and nowhere else on earth, so far as I have ascertained."[442] The fullest account is provided by Fr. Felix Fabri who, with Bernhard von Breydenbach[443] and Paul Walther,[444] visited Sinai in 1483. On September

---

[439] Antoninus (*Itinerarium*), 1849: p. 912 ("Inter Sina et Horeb est vallis in qua certis temporibus ros de coelo [cadit], quem manna appellant. Et coagulatur, et fit tanquam granum masticis; et habent inde plena dolia, et dant aliis pro benedictione, et nobis dederunt sextarios quinque; ex quo et perdite bibunt, et nobis dederunt bibere.") Cosmas Indicopleustes (6th century), merchant and later monk in Sinai, noted that Biblical manna "descended" in an area "half way between Elim and the Mount Sinai" (1897: p. 144).

[440] Georgius (*Chronographia*), 1652: pp. 128–129. For references to the manna of the Hebrews and to "food from heaven" in the Koran, see Bell, 1937: 1: p. 153; Pickhall, 1938: 1: II, 57, V, 113; 2: XX, 80.

[441] Frescobaldi, 1948: p. 58; Suriano, 1949: p. 144 ("like to that [manna] which the children of Israel ate, and I never tasted anything so sweet, pleasing and medicinal, and of no small price.") Konrad von Megenberg (*Das Buch der Natur*, 1349–1350) wrote "Vom Himmelbrot" (1897: p. 72), but without contemporary information from Sinai. I can find no reference to manna in Bartholomaeus Anglicus (fl. 1230–1250) *De Proprietatibus Rerum*, 1535.

[442] Harff, 1946: p. 140.

[443] Breydenbach (1486), 1502: n.p. (Fig. 11); 1905: p. 370.

[444] P. Walther (1892: pp. 200–220) does not appear to comment on manna; nor do the earlier or contemporary visitors Thetmarus (1217), 1851; Burchard (1280), 1896; Ludolphus de Suchen (1313–1350), 1895; Le Huen, 1488; Noe (1500), 1598. Ghistele ([1485], 1557: pp. 194, 277–278) refers to the manna of Egypt and Arabia, and of Persia and Cathay.

22nd the party came upon "manna or the dew of manna" in the upper part of the Wadi El Sheikh (on the route of the Exodus, and also most often mentioned by modern travellers). Fr. Felix remarked that, unlike the "miraculous" manna of the Old Testament, the natural product was to be found during only two months of the year (August and September) and "does not cover the surface of the earth, but hangs upon the leaves of plants and the points of stones, like dew."[445] He claimed to have "seen and eaten much of this manna"; that later bought at the monastery of St. Catherine was thought to be adulterated.

In the period between 1500 and 1800 many works in the fields of botany, pharmacology, and theology dealt with the problem of Sinaitic manna.[446] Almost without exception they broke little new ground. Pierre Belon du Mons (1555), among the small minority of writers to visit Sinai, confused the local product with alhagi manna (tereniabīn) and with "miel du cedre."[447] Prosper Alpinus in his De Medicina Aegyptiorum (1591) attempted to distinguish between tereniabīn and what he called terengibil, collected "supra arbores montis Synai in Arabia deserta."[448] In the works of Joannes Cotovicus (1619) and Gaspard Bauhin (1623) the manna of Sinai was firmly associated with agul (Alhagi sp.).[449]

Fr. Jean Thenaud travelled to Sinai in 1512 and on June 20th–22nd found manna "sur les arbres et roches."[450] Another Frenchman, Antoine Morison (1697), arriving in the Wadi Feirān in late November, could only repeat what he was told: that manna was collected in the early morning during the hottest part of the year (July and August), used by the local population in place of honey, and also sold to the monks of St. Catherine's.[451] That the Israelites were able to grind and bake their manna (Numbers 11.8) suggested to J. D. Michaélis the presence of broken leaves.[452] The first person to refer specifically to the tamarisk was, so far as is known, I. E. Fabri (ca. 1770) in his account De Manna Ebraeorum Opuscula.[453]

"It is from tarfa [tamarisk]" wrote J. L. Burckhardt, "that manna is obtained, and it is very strange that the fact should have remained unknown

---

[445] F. Fabri. 1892–1893: 2, 2: pp. 544–545. Breydenbach (doubtless following Fabri) and Arnold von Harff also mention August and September; similarly I. E. Fabri, 1776: p. 139. Antoine Morison ([1697], 1704: p. 90) gives July and August, and most 19th-century authorities refer to the period from late May to early July. These differences may be related to the localities examined and/or to variations in climate (notably rainfall) from one year to the next. Felix Fabri's route was discussed in detail by G. W. Murray, 1956: pp. 335–346.
[446] Brasavolus, 1537: p. 335; Palea and Bartholomaeus, 1550: p. 251; Lobel (1570–1571), 1576: pp. 23, 26; J. Bauhin (ca. 1600), 1650–1651: 1: pp. 188–190; J. Buxtorf (Dissertatio de Manna, ante 1629), 1747; M. Walther (Tractatu de Mannâ), 1633; Deusingius (Dissertationes de Manna et Saccharo), 1659; Johnstone, 1662: pp. 334, 338; Bochart, 1663: 2: p. 627; 1692: 3: pp. 59, 873 ff.; Pomet, 1694: 1: pp. 234–236.
[447] Belon du Mons, 1555: p. 129.
[448] Alpinus, 1591: p. 127.
[449] Cotovicus, 1619: p. 412; G. Bauhin (1623), 1671: p. 497. This was also the position adopted by Hallé in his account of agul in L'Encyclopédie Méthodique (1787: pp. 397–399).
[450] Thenaud, 1884: p. 70. Christoph Fürer ab Haimendorf ([1565–1566], 1621) and Henri Castela ([1600], 1603) also visited Sinai, but do not refer to manna.
[451] Morison, 1704: pp. 90–91.
[452] Michaélis, 1774: p. 109.
[453] I. E. Fabri, 1776: pp. 136–139 (manna tamariscina).

In hac valle in q̃ pꝛefatū eſt monaſteriū atcp etiã in alijs vallibꝫ p̃ cĩrcũitũ mõtis Synai.mãna in uenit in Auguſto ⁊ Septēbꝛi ðūtaꝛat.qð monachi colligētcꝫ ⁊ Arabeꝶ:aduētantibꝫ vendūt pegri nis. Ladit aũt verſus ðiē in modū roꝛis ⁊ pꝛuine.appendetcꝫ graminibꝫ guttatim.lapidibꝫ q̃ꝫ ⁊ ſo= lijꝶarboꝛ.cuncꝫ colligit in vnũ cõcurrit coagulū ſĩc piꝛ.⁊ ad ſolē vel igne reſoluit.Eſt autē guſta= tũ ſicut mel ðulce.ðentibꝫ adberēꝶ comedētis ðe eo multaꝶ cõparauim⁹ partes.ſ̃ꝫ ⁊ ðe ligno illo ðe q̃ erat ꝑ ga moyſi ſumpta:cũ qua táta fecit mirabilia in t̃ra egypti in ðecē plagis.Poꝛro in ĩpo mo= naſterio grãdis qdam eſt ciſterna in petra inciſa.aqꝶ ſemp qſi miraculoſe abūdanꝶ.Et ðicũt monꝛ chi meriũꝶ moyſi illas ſe aqꝶ b̃ꝛe.Jnde oĩa vaſa nr̃a impleuiu⁹ vt p̃ ðeſertũ aquas haberem⁹.q̃a iã tꝑus imminebat nr̃i receſſuꝶ Arabeꝶ frequēter iuꝛta ðictũ ſunt monaſteriũ.⁊ mirũ ĩmodũ ipſum grauãt.ita cp ðietim abbas ciuſdē monaſterij octoginta vel centũ arabes in panibꝫ ⁊ pultibꝫ paſcere cogit Jn eodē monaſterio omniũ nationũ ⁊ ſectaꝛ. pſone recipiunt ad oꝛdinē.erceptiꝶ ſolis Arme nis et Jacobitis.recepti autē moꝛe coꝛũ viuere tenēt.et inſtituta ſeruare grecoꝛum.

Fig. 11. Account of the manna of Sinai. Bernhard von Breydenbach, 1483–1484 (1502: no pagination).

in Europe till Seetzen mentioned it in a brief notice of his tour to Sinai (1807) . . . ."[454] U. J. Seetzen referred particularly to the tamarisk thickets of the Gharandal, Feirān and El Sheikh, and to the Wadi Taib where he found branches coated with manna.[455] Burckhardt's own enquiries date from May 1816 when he was travelling through the Wadi El Sheikh:

"In the month of June [manna] drops from the thorns (sic) of the tamarisk upon the fallen twigs, leaves, and thorns . . . .; the manna is collected before sunrise, when it is coagulated, but it dissolves as soon as the sun shines upon it. The Arabs clean away the leaves, dirt, etc which adhere to it, boil it, strain it through a coarse piece of cloth, and put it into leathern skins; in this way they preserve it till the following year, and use it as they do honey, to pour over their unleavened bread, or to dip their bread into. I could not learn that they ever make it into cakes or loaves. The manna is found only in years when copious rains have fallen; sometimes it is not produced at all, as will probably happen this year. I saw none of it among the Arabs, but I obtained a small piece of last year's produce in the convent [of St. Catherine] . . . In the season at which the Arabs gather it, it never acquires the state of hardness which will allow of its being pounded . . . . Its colour is a dirty yellow, and the piece which I saw was still mixed with bits of tamarisk leaves; its taste is agreeable, somewhat aromatic, and as sweet as honey."

"The quantity of manna collected at present, even in the seasons when the most copious rains fall, is very trifling, perhaps not amounting to more than five or six hundred pounds. It is entirely consumed among the Bedouins who consider it the greatest dainty which their country affords. The harvest is usually in June, and lasts for about six weeks; sometimes it begins in May."[456]

The Wadi El Sheikh is notable for almost pure stands of *T. mannifera*;

[454]Burckhardt, 1822: pp. 599–600. Karsten Niebuhr ([1761–1764], 1792: 2: p. 360) admitted that "we neglected to inform ourselves, in Arabia, concerning the production of manna . . . ."

[455] Seetzen, 1808: p. 151; 1854–1859: 3: pp. 75–79, 129. Seetzen considered that the manna of the Old Testament may have included the edible exudations of both tamarisk and acacia (gum).

[456] Burckhardt, 1822: pp. 600–601. J. R. Wellsted (1838: 2: p. 51) put the amount collected each year at no more than 700 pounds.

indeed the upper part is known as *Wadi Tarfā*. It was here in June 1832 that N. Bové saw manna being collected by women and children, and observed that "les Arabes clarifient cette manne en la dissolvant dans l'eau chaude, et en écumant cette espèce de sirop."[457] The substance was highly valued for its saccharine quality[458] and could be stored for several years in gourds, skins, and casks.[459] A. W. L. Lindsay (1837) noted that it was used to sweeten bad water.[460] Small quantities appear to have been traded as far as Cairo (probably for medicinal rather than alimentary purposes), but the chief commercial outlet was the monastery of St. Catherine and thence to visiting pilgrims.[461] F. S. Bodenheimer calculated that a man could collect about a kilogram in a day at the height of the season.[462] However the amount available varied greatly from year to year and between one *wadi* and the next in the same year. Sometimes several years passed with little or no harvest. The Feirān drainage system (including El Sheikh) broadly corresponds to the central zone of manna production,[463] which extends as far north as the Wadi Gharandal[464] and as far south and east as the Isla and the Nasb.[465] According to Ritter (1848), "the manna-producing [tamarisk] cannot grow at an elevation of 3000 feet above the sea .... Nor does this tree flourish and yield its gum in the extremely dry regions of the peninsula."[466]

Important new light was shed on the question of the uneven distribution of tamarisk manna by C. G. Ehrenberg's discovery (ca. 1823) of a parasitic scale insect – named by him *Coccus manniparus* – in the Wadi Esle (Isla).[467] Ehrenberg assumed that the manna flowed from punctures made by the coccus in the tender outer twigs of *T. gallica* var. *mannifera*. This explanation was widely accepted, although several later travellers failed to find the insect in association with manna.[468] F. S. Bodenheimer re-investigated the matter in 1927 and, on the basis of intensive field observations, came to the conclusion that tamarisk manna was not an exudation, occasioned by

---

[457] Bové, 1834: p. 166. The manna of the Wadi El Sheikh is also mentioned by Tischendorf (1844), 1862: p. 54; E. H. Palmer, 1871: p. 81; Ebers, 1872: pp. 223–224; and possibly also by Fazakerley (1811), 1820: p. 376.

[458] On the chemical composition of Sinaitic manna (55 per cent cane sugar), see Berthelot, 1863: pp. 82–85; Fodor and Cohn in Bodenheimer and Theodor, 1929: p. 89.

[459] Lindsay, 1838: 1: p. 311; Wellsted, 1838: 2: p. 51.

[460] Lindsay, 1838: 1: p. 310.

[461] Henniker, 1823: p. 228; Stephens, 1838: 1: p. 315; Robinson and Smith (1838), 1841: 1: p. 170; Brockbank, ca. 1920: p. 23.

[462] Bodenheimer, 1947: pp. 2–3.

[463] Morison (1697), 1704: pp. 90–91; Rüppell, 1829: p. 190; Lepsius, 1846: p. 66; Maughan, 1873: p. 99.

[464] Lindsay, 1838: 1: p. 311.

[465] Burckhardt, 1822: p. 601; Bodenheimer, 1947: p. 2.

[466] Ritter, 1866: 1: p. 271.

[467] Ehrenberg, 1827a: pp. 241–282; 1827b: pp. 68–78; Ehrenberg and Hemprich (1820–1825), 1900: p. 1; Anon., 1828: p. 262; Bach, 1857: p. 289. Cf. Forskål, 1775b: p. XXIII (*Cicada mannifica*). According to Ritter (1822–1859: 14: p. 672; 1866: 1: p. 276), the Wadi Esle is known to the local Bedouin as Ain el Man, "fons mannae." H. C. Hart (1891: p. 22) reported that "At the head of Wadi el Ain, a grove of tamarisks was plentifully indued with an excrescence or exudation of grayish-white pilules of a viscid substance, with a faint taste of nucatine."

[468] Wellsted, 1838: 2: p. 51 (September); Tischendorf, 1862: pp. 54–62 (May to June).

punctures, but an excretion or "honeydew" of two related parasites, *Trabutina mannipara* (the *Coccus manniparus* of Ehrenberg) and *Najacoccus serpentinus* var. *minor* Green,[469] the former found in the highlands and the latter in the lowlands of Sinai. This gives point to P. Haupt's contention (1922) that "the primary connotation of Hebrew *man*, manna, is not gift, but separation, elimination, secretion; it denotes also the manna insect...."[470]

Much has been written on the nature and origin of Hebraic manna. The geographer Karl Ritter argued the case for the tamarisk at some length.[471] The details found in the modern versions of Exodus and Numbers can only be partly matched by natural phenomena. Bodenheimer however maintained that the earliest Scriptural accounts[472] give a "remarkably suitable description" of the manna collected today on or beneath the branches of the tamarisk. He also rightly emphasises that the Israelites collected their manna at the appropriate time of the year (May and June) and in precisely the area (Elim to Rephidim [Feirān]) where the tamarisk product is known today. "Honeydew" is excreted during the day (attracting bees), falls or accumulates at night (thus manna as "dew," from the sky or heaven), and in the morning it is often consumed by ants (a further connection with insects or "worms"). From our knowledge of the dietary preferences of desert peoples, we can assume that such a saccharine substance would have been regarded as a luxury and, very probably, its importance exaggerated. No case that is as strong can be made out for other known mannas. In particular, the lichen *Lecanora esculenta* (*supra*, pp. 43–54), while fitting the Biblical account in some respects,[473] has never been reported from Sinai, either *in situ* or as a wind-borne deposit.[474]

---

[469] Bodenheimer in Bodenheimer and Theodor, 1929: pp. 45–88 (the synomyms of *Trabutina mannipara* [Ehrenberg] Bdhmr. are listed on p. 64). See also Bodenheimer, 1928–1929: 2: p. 301; 1937: p. 220; 1947: pp. 2–6; Grassé, 1949–1951: 2: p. 1646. Much earlier, R. Blanchard (1883: p. 67) questioned whether the tamarisk manna of Sinai might not be an insect excretion.

[470] Haupt, 1922: p. 235. According to Bodenheimer (1947: p. 6) "*man* is the common Arabic word for plant lice."

[471] Ritter, 1822–1859: 14: pp. 665–695; 1866: 1: pp. 271–292. See also Büsching, 1775: pp. 41–48; Virey, 1818: pp. 120–126; Raumer, 1837: pp. 26–28; Schubert, 1838–1839: 2: pp. 346–349; Rosenmüller (1830), 1840: pp. 320–331; Langerke, 1844: pp. 444–450; Hogg, 1849: pp. 183–236; Bonar, 1857: pp. 146–155; James, 1872: pp. 59–62; Ebers 1872: pp. 223–234; Kolb, 1892: pp. 1–13; Teesdale, 1897: pp. 229–233; Bourgon, 1898: pp. 41–42; Petrie, 1906: pp. 230–231; Pilter, 1917: pp. 155–206; Moghadam, 1930: pp. 123–135.

[472] On the composite nature of the manna tradition, see Coppens, 1960: pp. 473–489; Borgen, 1965; Malina, 1968.

[473] O'Rorke, 1860: pp. 412–419; Renard and Lacour, 1880: pp. 3–20. W. T. Pilter (1917: p. 205) and P. Haupt (1922: p. 235) thought that lichen manna and tamarisk manna were probably combined (for purposes of grinding and baking).

[474] Holmes (1920: p. 175) argued that the characteristics described "belonged to fungi rather than to lichens." Moldenke (1952: p. 126) proposed "three distinct types of manna": *Tamarix* sp. and/or *Alhagi* sp., the algal genus *Nostoc*, and species of *Lecanora*. T. L. Phipson (1856–1857: p. 530) reported "an efflorescence of mannite" on marine algae. Lady Anne Blunt (1879: 2: pp. 21–22) mentions "a suggestion that the manna .... [of] .... the wilderness" consisted of truffles (*kemeyes*; cf. Burckhardt, 1830–1831: 1: pp. 60–62 [*kemmáye, kemmá*]). Related to this we have T. von Heldreich's opinion (1862: p. 6) that the tuber *Cyperus esculentus* L., known as "manna" in Greece but introduced from Egypt, was Hebrew manna. This may help to explain references to "manna" in Greece (*supra*, p. 13 n. 3). For *C. esculentus* in Egypt, see Muschler, 1912: 1: p. 173.

J. MINOR SOURCES OF MANNA

(a) Astragalus spp.

A. Haussknecht (1870) maintained that manna was collected from two closely related species of *Astragalus, adscendens* Boiss. et Haussk. and *florulentus* Boiss. et Haussk.[475] (Map 13).[476] The "best kind" was known as *ges alefi* or *ges chonsari*. However these descriptions appear to refer to oak manna and tamarisk manna respectively, also collected in southern and western Persia.[477]

Astragalus adscendens is one of several species that exude gum tragacanth, either naturally or following artificial incisions.[478] Available between Asia Minor and Afghanistan, the gum was known to the Greek physicians and may originally have been valued as food.[479] A third species, *A. fasiculaefolius* Boiss. yields a substance, sarcocolla, that is "sharp and sweetish, followed by a nauseous and disagreeable bitterness."[480] Sarcocolla found a place in the *materia medica* of the Arabs who recommended that it should be applied to wounds. Never, so far as is known, was it described as "manna."

Tragacanth and sarcocolla may have been used to adulterate substances marketed as *gaz-angubīn*. The product of *A. florulentus*, on the other hand, appears to have been regarded as a true manna,[481] although no special name has been found. M. Meyerhof (1940) reported that the manna of *Atraphaxis spinosa* (*šīr-ḫušk*) and of *Astragalus* sp. "sont en vente dans les bazars du Caire sous le nom de *mann fārsī* (*manne persane*) .... La manne des astragales est moins blanche et moins bonne."[482]

(b) Cedrus sp.

Hippocrates (ca. 460–377 B. C.) refers to *mel cedrinum*, added to wine, in a prescription for ulcers (*De ulceribus*).[483] Perhaps this is the *roscidum mel*

---

[475]Haussknecht, 1870: p. 246.

[476] Boissier, 1867–1888: 2: p. 317; Sabeti, 1966: no. 99; Guest and Townsend, 1966–1974: 3: pp. 326–328. Wood and Bache (1907: p. 764) add *A. anisacanthus* Boiss. (Khunsar, Feridan, Chahar Mahal, and Ispahan). *Astragalus* (Leguminosae) includes upwards of 2000 species (shrubs and herbs).

[477] *Chonsar*, from the town of Khunsar, about 150 kilometres north west of Ispahan (not to the south west, as in Haussknecht). Haussknecht's statement has been quoted or followed by several later authorities: Ludwig, 1870: p. 35; Flückiger and Hanbury, 1879: p. 415; Flückiger, 1883: p. 27; Dymock, 1890–1893: p. 161; Hooper, 1909: p. 33; Laufer, 1919: p. 348; Andreu, 1953–1955: 13, 29: p. 200. Moghadam (1930: p. 140) recognised the error and did not include *Astragalus* among the producers of manna in Persia.

[478] Boissier, 1867–1888: 2: p. 317 ("gummi copiosum praebet"); Tease, 1936: pp. 206–208; Howes, 1949: pp. 35–51. See also P. Schwarz, 1896–1936: 5: p. 627, n. 5.

[479] Guest and Townsend, 1966–1974: 3: pp. 326–328.

[480] Hooper, 1913: pp. 177–181; 1931: p. 306. See also Polak, 1865: 2: p. 285.

[481] Dragendorff, 1898: p. 322; Guest and Townsend, 1966–1974: 3: pp. 234–235 ("flake manna," *gazanjabín*). Cf. Ritter, 1822–1859: 14: p. 689; Achundow in Muwaffiq ibn ʿAlī, 1968: p. 355 (*gezengebín*); Wehmer, 1929: p. 347 (*gesengebin*). I have found no illustration of *A. florulentus*.

[482] Meyerhof in Maimonides, 1940: p. 194.

[483] Hippocrates, 1825–1827: 3: p. 316. Fothergill (1746: p. 92), following Fuchsius (ca. 1550, unlocated) suggested that perhaps a comma had been omitted and that two substances were implied, "honey" and "cedar resin."

from Mount Lebanon mentioned by Claudius Galenus (ca. A.D. 129–200) as something of a curiosity.[484] It is not clear that either physician had seen the product. Later commentaries were confused by the fact that the cedar of Lebanon and the Taurus (*Pinus cedrus* L., *Cedrus libani* Barrel, *C. libanotica* Link) yields a resin (*cedria*),[485] usually from artificial incisions in the trunk, and a sweet manna (*mel*), naturally but infrequently and in comparatively small quantities, from the leaves and possibly the branches.[486] Moreover, there is a further possibility of confusion. *Manna* (in the sense of "grain") *libani*[487] usually refers to the gum-resin *olibanum* (frankincense), the Greek *libanos* and Hebrew *lebonah*, a product not of the cedar and of Lebanon, but of *Boswellia* spp. of southern Arabia and Somalia. This was more often described in Classical and later sources as *manna thuris* (Latin *tus* or *thus*, incense; *supra*, p. 7).

Early modern accounts appear to be largely based on two works by Pierre Belon du Mons.[488] Unfortunately Belon does not clearly distinguish between "miel de cedre" (of Hippocrates) and two other mannas, "tereniabin" (*tar-angubīn*, obtained in Persia and central Asia from *Alhagi maurorum* and *A. camelorum*), and the product collected in Sinai and elsewhere from *Tamarix* spp.

Johann Bauhin (died 1613) included *mel libani* and *mel cedrinum* in a substantial description of various mannas,[489] and likewise his younger brother, Gaspard (died 1624).[490] Pierre Pomet (1694) noted the rare and esteemed "gomme de cedre [du Liban] ou manne masticine . . . . qui est par grains comme le mastic, d'ou est venu son surnon."[491] Here the resin and the manna again appear to be confused. Moreover, John Fothergill (1746) was probably correct in suggesting that *manna mastichina* [*orientalis*] was not "cedrine manna" but *tar-angubīn*.[492] According to X. Landerer (1854) "*manna cedrina* appears in small globules on the branches of *Pinus cedrus*. It is brought from Mount Lebanon where a very small quantity of 2 or 3 *drachms* fetches from 30 to 40 *piastres* [10 to 13 shillings]. In Syria it enjoys considerable reputation in the *Marás* or phthisis [pulmonary tuberculosis], and it is an ingredient in electuaries [medicinal powders] . . . ."[493]

---

[484] Galenus, 1530: p. 106.
[485] Theophrastus, 1961–1968: 2: p. 225. Cf. Loiseleur-Deslongchamps, 1837: p. 47.
[486] Hare *et al.*, 1905: p. 953. In addition, an oil (*cedrium*) was distilled from the wood of *C. libani* (Pliny, 1961–1966: 7: pp. 14–15, *cedri sucus*; 4: pp. 420–421, *cedrium*).
[487] For example, Paulus Aegineta (ca. 640), 1914: p. 683.
[488] Belon du Mons, 1553: pp. 8b–10b (*cedrinum mel*); 1555: p. 129. The only earlier reference found is in Brasavolus, 1537: p. 336. See also *Ricettario Fiorentino*, 1567: p. 44; Lobel (1570–1571), 1576: p. 24.
[489] J. Bauhin, 1650–1651: 1: pp. 180, 183, 190–191.
[490] G. Bauhin, 1671: p. 497 (*cedrium mel* [Hippocrates], *ros libani* [Graecis], *terniabin* [Arabibus et Turcis]). Cf. Johnstone, 1662: p. 334.
[491] Pomet, 1694: 1: pp. 116 (6 oz. per day from the trunk and branches at times of high atmospheric temperature), 238; *tereniabin* he considers separately (p. 239). Geoffroy ([died 1731], 1741: 2: p. 590) noted *mastichina* among the Oriental mannas. Savary des Bruslons (1742: 2: pp. 715, 1185) followed Pomet. I. E. Fabri (1776: p. 104) gives *manna cedrina* (Hippocrates).
[492] Fothergill, 1746: p. 90. Cf. Belon du Mons, 1553: p. 10a; J. Bauhin, 1650–1651: 1: pp. 186, 198–199; Flückiger, 1883: p. 25 (*manna granulata* or *mastichina*).
[493] Landerer, 1854: pp. 411–412; Simmonds, 1895: p. 135.

(c) Dendrocalamus sp.

In March 1900, A. E. Lowrie, a forest officer working in the Central Provinces of India, observed a sugary deposit on the common bamboo (*Dendrocalamus strictus* Nees). "This .... extended for about 5 feet along the culms and was entirely absent towards the tops; it was found both at the nodes of the bamboo as well as on the stems between the nodes .... this has nothing to do with any insect deposit, nor has it been caused through the aid of insect punctures .... The culms .... were one, two, and three years old."[494] At a time of general scarcity the manna was "collected by the

Map 13. Combined distribution of *Astragalus adscendens* Boiss. et Haussk., *A. florulentus* Boiss. et Haussk., and *A. fasciculaefolius* Boiss.

---

[494] Quoted by Hooper, 1900: pp. 187–188.

handfulls." D. Hooper found that it did not contain mannite, but rather "a saccharose (95.63 per cent), related to, if not identical with, cane-sugar .... and might be used for cooking or making sweetmeats in the place of ordinary sugar."[495]

(d) Indigofera sp.

A manna-producing species, known in the Aïr region of north-central Africa as *tagjao* or *tajjaoua*, has been identified as *Indigofera semitrijuga* Forsk. A. Chevalier received specimens from M. le capitaine Le Rumeur who reported that "elle secrète une matière sucrée qui est mangée par les indigènes [un groupe nomade]. Le sucre coule à la base et parfois forme bloc avec le sable."[496]

(e) Morus sp.

Francis Bacon (1561–1626) in his *Sylva Sylvarum* maintained that the manna of Calabria was found on the leaves of the mulberry (*Morus alba* L.) and "not upon other trees."[497] As to the latter, he was certainly misinformed. Concerning the mulberry, Sadegh Moghadam has the following note:
"*Manne de Mûrier*: nous avons pu remarquer sur des mûriers blancs au cours de notre voyage en Perse, en 1928, aux environs de Téhéran, qu'une exsudation sucrée brunâtre s'écoulait en grande quantité pendant le mois de septembre. Cette manne se produit au cours des étés chauds; elle est négligée et inutilisable."[498]

(f) Olea sp.

The earliest known reports of manna belong to the 18th century.[499] L. de Luca (1863) found mannite in "nearly all parts" of the olive (*Olea europaea* L.).[500] Later J. A. Battandier observed and analysed an exudation (52 per cent mannite) on trees in gardens at Mansourah, 150 kilometres south east of Algiers.[501]

---

[495] S. G. Harrison (1951: p. 416) noted that "A sugary exudation previously reported on *Bambusa arundinacea* Willd. and "*Bambusa verticellata*" has never been confirmed and may have been confused with *tabashir*, a siliceous substance which is produced inside the stems of certain bamboos, and has long been used in Indian medicine ...."
[496] Chevalier, 1933: pp. 276–277, 281. The plant was at first thought to be *Alhagi* sp., another member of the Leguminosae. *I. semitrijuga* is found between Arabia and Mauritania.
[497] Bacon, 1627: p. 203.
[498] Moghadam, 1930: p. 29. See also Targioni-Tozzetti (1715), 1768–1779: 6: pp. 423–424.
[499] Geoffroy, 1741: 2: p. 590; Sestini, 1788: p. 92.
[500] Luca, 1863: p. 473.
[501] Battandier, 1901: p. 117. Subsequent reports appear to be based on Battandier.

(g) Palmeae

It has been suggested that the substance known to Dioscorides and Pliny as *elaiomeli* and found on the trunks of an unidentified tree in parts of Syria was "manna" exuded by species of palm.[502] Arab authors provide more specific information for the eastern Maghreb. Ibn al-Baiṭār (1197–1248) and 'Abd ar-Razzāq of Algiers (18th century) quote Ibn al-Djazzār, a physician of Qairwan, to the effect that manna "falls on the branches of the palm" in *Kastîliya (Casthilya)*, southern Tunisia.[503] This is also the region, around Tozeur (Map 8), where manna was collected according to the geographer al-Bakrī (ca. 1040–1094).[504] M. Gast (1968) reported that in the Ahaggar "les palmiers dattiers [*Phoenix dactylifera* L.] .... exsudent des gouttelettes sucrées entre les dattes sur les branches des régimes."[505]

(h) Pinus spp.

Along the foothills of the western Himalaya, notably in Kumaun and the region of Simla, *Pinus excelsa* Wall. ex Lamb. (Bhutan pine) occasionally yields a sweet edible manna. This was first reported by W. Madden (1850).[506] G. Watt (1890) observed that "As a probable consequence of an exceptionally dry autumn the pines of the western Himalaya [*P. excelsa* and *P. longifolia* Roxb.] have been exuding manna from the tips of the twigs .... It is not reputed to be used medicinally, but is collected and eaten, or employed in adulterating honey."[507]

(i) Platanus sp.

E. Jandrier (1893) noted: "Pendant les étés secs, on peut recueillir sur certains platanes [*Platanus orientalis* L.] une exsudation de consistance et d'aspect variables, tantôt sèche et brillante, tantôt pâteuse et jaunâtre, renfermant, à côté d'une faible quantité d'un sucre réducteur paraissant être

---

[502] Dioscorides (ca. A.D. 78), 1952–1959: 3: p. 35 (near Palmyra; comments by Andrés de Laguna); Pliny (A.D. 23–79), 1961–1968: 4: pp. 310–311 ("There is an oil that grows of its own accord in the coastal parts of Syria called *elaeomeli*. It is a rich oil that trickles from trees, of a substance thicker than honey but thinner than resin, and having a sweet flavour; this also is used by the doctors." Cf. Mela (1st century A.D.), 1967: p. 70 (India). See Watt, 1889–1893:5: p. 165; Forbes, 1966: p. 100.

[503] Ibn al-Baiṭār, 1877–1883: 1: p. 309, no. 408; 'Abd ar-Razzāq, 1874: pp. 342–343, no. 876. For Ibn al-Djazzār and other members of the school of Qairwan, see Ibn Milad, 1933: pp. 26–47.

[504] Al-Bakrī, 1913: p. 104, no. 118. See also Renaud and Colin (*Tuḥfat al-aḥbāb*), 1934: p. 116.

[505] Gast, 1968: p. 243. The "sap" of various palms (for example, *Caryota urens* L. *Borassus flabellifer* L.) is boiled to provide sugar ("jaggery") and palm wine or "toddy" (Tennent, 1860: 1: p. 112; 2: p. 524).

[506] Madden, 1850: pp. 8–11. See also Brandis, 1874: p. 512; Flückiger, 1883: p. 28; Dey, 1896: p. 113; Henry, 1924: p. 389 (? aphid origin).

[507] Watt, 1889–1893: 3: p. 443. *P. excelsa* = *P. wallichiana* A. B. Jacks; *P. longifolia* = *P. roxburghii* Sarg. (Chir pine).

de la glucose, de 80 à 90 pour 100 de mannite, qu'on peut extraire avec la plus grande facilité, par cristallisation, de l'alcool bouillant."[508]

(j) Salsola sp.

There are at least two independent reports of manna on the halophytic herb *Salsola foetida* Del. (Arabic *mulleyh*).[509] J. E. T. Aitchison wrote: "At Sha-ishmail [Shāh Ismāīl, south west Afghanistan], on the 28th October 1884, I obtained from the surface of [the] leaves a quantity of manna, which presented the appearance of drops of milk that had hardened on its foliage; this seemed to be well known to the Baluchi camel drivers, who collected and ate it. The only name they had for the substance was *shakar* (sugar)."[510] M. Gast found that in the Ahaggar "*Salsola foetida* Del. ("issin"), *Publicaria crispa* L. ("tanetfert"), *Altriplex halimus* L. ("aramas") fournissent parfois aussi un peu de manne; seuls les enfants s'amusent à ramasser ces menues récoltes."[511]

---

[508] Jandrier, 1893: p. 498.
[509] Délile, 1812: p. 57. Illustrated in Barbey, 1882: tab. 8, fig. 11.
[510] Aitchison, 1891: p. 181; 1886–1887: p. 467 ("aromatic manna").
[511] Gast, 1968: p. 243. *Publicaria crispa* = *Pulicaria crispa* Sch. *A. halimus* L. is the "Mediterranean saltbush." *S. foetida* serves as fodder for camels.

4 MANNAS OF EUROPE

A. CISTUS sp.

Around the shores of the Mediterranean several species of *Cistus* exude a fragrant substance known as labdanum or ladanum (Hebrew *lōt*, Persian *lād, lādan*; Latin *leda, lada*, "gum cistus").[1] Labdanum has been employed from ancient times to the present day in the preparation of perfumes and incense. Spanish labdanum, *ládano*, is obtained from *C. ladaniferus* L.,[2] which also yields a sweet manna.[3] This was investigated by the Royal College of Physicians of Madrid in 1752.[4] Apparently the manna had been put to little or no use before the middle of the 18th century.

Joseph Quer in his monumental *Flora Española* observed that *C. ladaniferus* was the source of three substances, including *manna de España* (locally *mangla*).[5] The English traveller J. T. Dillon (1778–1780) found that in the neighbourhood of Burgos "the old branches [of the gum-bearing cistus] distil a liquid matter which the heat of the sun condenses into a white sugary substance . . . . a true manna; it is gathered and eaten greedily by shepherds and boys . . . . [and is] equal in goodness to that of Calabria" (the manna of *Fraxinus ornus*).[6] *C. ladaniferus* (*jara, jara común, jara negra, jara pegajosa*) is widely distributed in Spain,[7] but is found more particularly on the slopes of the Sierra Morena in Andalucia. "Spanish manna" was not exploited commercially on any considerable scale and it never achieved the reputation of that from Calabria and Sicily.

---

[1] Howes, 1949; p. 158. On the name *ladanum*, see André, 1956; p. 177.

[2] Lázaro é Ibiza, 1906–1907: 2: p. 269.

[3] *C. ladaniferus* is the *Cistus Ledon* of Clusius (1576: pp. 155–169), but he makes no reference to manna. Landerer (1854: p. 412) refers to "manna cistina sive labdanifera" from Greece. Goetz (1888–1923: 3: pp. 591, 613, 625) gives *ladanus*, "mel in folio ulmi."

[4] Dillon, 1780: pp. 127–128; Proust, 1806: pp. 144–145. Collected by the botanists Cristóbal Velez and Juan Minuart (Mas y Guindal, 1953: p. 81). See also Andreu, 1953–1955: 15, 38: p. 313.

[5] Quer, 1762–1784: 4: p. 325. Cf. Krünitz, 1808–1828: 83: p. 740. The *Diccionario* of the Real Academia Española (ed. 1970) gives *mangla*, "resina de la jara [*Cistus*], ládano" in the Sierra Morena. *Magna'* (manna) is mentioned in a list of "medical recipes" in Judeo-Spanish ca. 1600 (Crews, 1967: p. 220, no. 15).

[6] Dillon, 1780: pp. 127–128 (according to Dillon, Spain could have supplied the whole of Europe). Noted by Woodville, 1790–1794: 1: p. 105; Flückiger and Hanbury, 1879: p. 416.

[7] Laguna y Villanueva, 1883–1890: 2: pp. 418–419. Also in Portugal, southern France, and North Africa (Boissier, 1839–1845: 2: p. 60).

B. FRAXINUS spp.

The most celebrated official manna of Europe was that found on species of ash, notably the "flowering ash," *Fraxinus ornus* L. *F. ornus* is highly variable when reproduced from seed and may properly include *F. rotundifolia* Miller,[8] a description more often applied to cultivated forms. The "common ash," *F. excelsior* L., also yields manna, but less abundantly. *F. ornus* is a native of the lands of the northern and eastern Mediterranean (Map 14).[9] In Italy the species is known as *frassino, fioriti, orno, ornello* or *avornello* (locally *amollei*); in Sicily, *frascinu di manna* (locally *muddia* or *middia*). *F. rotundifolia* (*F. ornus* var. *rotundifolia*) has been reported as far east as peninsular India.

The production of manna was largely confined to central and southern Italy and to Sicily (Map 15). Isolated reports exist for southern France and Spain.[10] In the foothills of the Himalaya, *F. excelsior* and *F. floribunda* Wall. yield manna (*shirkhist*) after incisions in the bark, but this, so far as is known, has not been collected on any significant scale or incorporated in the local *materia medica*.[11]

(a) Fifteenth to seventeenth centuries

What became the principal medicinal manna of Europe appears to have been unknown in Classical times, notwithstanding the Greek name for *Fraxinus ornus*, μελία. Nor, apparently, do the well-informed Arab physicians describe the ash manna of southern Europe. That it may have been collected during the Saracenic occupation of Sicily (827–1070) is however suggested by the name *Gibilmanna* (*Jabal Mann*, "manna mountain"),[12] south of Cefalù. This was one of the chief producing areas at a much later

---

[8] Cleghorn, 1870: p. 132; Hanbury, 1876: p. 367.
[9] North to southern Germany and Switzerland, and more widely as an ornamental. It was (re-) introduced to England ca. 1730. A. E. Hunter (1969: p. 218) described the sweet exudate of *F. ornus* "from the knotted wood of the petrified forests which covered the [Liverpool] area in remote times." There are similar "reefs" at Tamerton and Dunchideock in South Devon.
[10] Duhamel du Monceau, 1758: 1: 152 (the valley of Fresquet, near Pennautier, in the exceptionally hot and dry summer of 1754); Merat and Lens, 1829–1834: 4: p. 221. A. Russell ([1756] 1794: p. 266) noted *F. ornus* (but not manna) in the vicinity of Aleppo. Cf. Ducros, 1930: p. 58 (*F. excelsior*, "derdar," "shagar el mann"). Löw (1967: 2: p. 286) gives "dardar."
[11] Royle, 1839: 1: p. 266; Watt, 1889–1893: 3: pp. 440–441; Dey, 1896: p. 113; Bamber, 1916: p. 7; Kirtikar and Basu, 1918: 2: pp. 768–769 (implying limited export); Chopra, 1933: p. 491; S. G. Harrison, 1951: p. 414. Harrison refers to "a specimen of manna [at Kew] from *F. ornus* var. *rotundifolia* from Madras .... [also] *shirkhist* from *F. floribunda* Wall. from Lahore, and a manna from a species of *Fraxinus* in Herāt (Afghanistan)...." Bunge (1835: p. 135, no. 343) reported *F. floribunda* from China/Mongolia. See also Franchet, 1883–1888 6: p. 83.
[12] Amico e Statella, 1757–1760: 3: p. 242 ("Gibilmanna," *Mons Mannae*). The name first appears in a document of 1082. According to Wenrich (1845: pp. 290, 318), the manna ash, as well as sugar cane and cotton, were brought to Sicily by the Arabs. Forbes (1966: p. 101) claimed that the manna of the ash "is .... not mentioned before the ninth century A.D. in Venetian bills for products imported from Sicily and Calabria"; no documentation is cited and no support for the statement has been found.

date. The manna available in Messina in the early 14th century[13] could have been of local origin, but more likely came from the Levant.[14]

The earliest direct reference to the manna of southern Italy is in Saladinus di Ascoli's *Compendium Aromatariorum*, written about 1430, and "the first work in world literature composed for apothecaries only."[15] This includes a calendar indicating the most suitable dates for the collection of herbal specimens, and under May we find the "manna of Calabria."[16] Exploitation may have commenced (or recommenced) about this time, for the product is not mentioned in Antonio da Uzzano's *Libro di Gabelle* (ca. 1442) which has sections on Naples and Calabria, Apulia, and Sicily (Palermo and Messina).[17] Raffaello Maffei [Volaterranus] implied that manna was first collected in Calabria in living memory (mid to late 15th century) and that it was considered inferior to the Oriental variety.[18] Jovianus Pontanus (1426–1503) described in poetic form the collection of manna in the valley of the Crati, and his words suggest a partial awareness that the honey-like substance was an exudation and not some kind of atmospheric dew.[19]

Antonius Musa Brasavolus (1537) of Ferrera maintained that there were three kinds of manna "in Calabro solo collecta": that found on the leaf (the best), secondly on the trunk, and, of least value, on the ground (*manna terrae*).[20] Matthiolus (1544) distinguished between the product of "Arabia" (*tereniabin*) and that of Calabria and Apulia.[21] Leandro Alberti's *Descrittione di tutta Italia* (1550–1551) likewise mentions Calabria and Magna Graecia,[22] but not Sicily. Two Franciscan fathers were the first clearly to establish (by protecting *orni* from the night air) that ash manna was an exudation; their observations were published (1550) in a commentary on the work of the great Arab physician Ibn Māsawaih (777–857).[23] Within a decade we have

---

[13] Pegolotti, 1936: p. 109.

[14] Similarly the manna discussed by Johannes [Mattheus] Platearius (12th century) of Salerno in *Liber de Simplici Medicina*, 1524: p. XXV; 1913: pp. 114–115; ca. 1972: pp. 188, 190–191. Guillaumin (1946: p. 199) states, without authority, that manna was collected in Sicily during the Middle Ages ("Au XIIIe siècle, la production de la manne subit un déclin en Sicile qui en était le principal pays producteur mais l'exploitation avait repris au XVe siècle et subsista jusqu'au début du XIXe siècle; . . . .")

[15] Muntner in Saladinus, 1953: p. iii.

[16] *Liber Saladini* in Ibn Māsawaih, 1502: p. 350 a; 1581: 2: 257 (earlier editions of the *Opera*, from 1491, also include Saladinus's treatise). The first Latin edition of the *Compendium* appeared in Bologna in 1488. It may have been composed in Hebrew. Saladinus (Salah-el-Din), a Jew, was court physician to the Duke of Taranto.

[17] Antonio da Uzzano, 1766: pp. 96–98, 164, 165, 169, 193, 196, 197.

[18] Maffei (1451–1522), 1559: p. 915 ("Manna, nostra ætate cœpit in Calabria provenire: licet orientali inferior.") Earlier editions of the *Commentatiorum urbanorum* were published in Rome (1506) and Paris (1511, 1515).

[19] Pontanus, 1513: p. 113 (*De Pruina et Rore et Manna*). Both Maffei and Pontanus are mentioned by Fiore da Cropani, 1691–1743: 1: p. 253.

[20] Brasavolus, 1537: p. 335.

[21] Matthiolus, 1544: p. 48; 1558: pp. 74, 244; similarly Laguna (1570) in Dioscorides, 1952–1959: 3: pp. 176–177.

[22] Alberti, 1551: pp. 171, 183.

[23] Palea et Bartholomaeus, 1550: p. 252. References are made to earlier editions (1543, 1545) which I have been unable to consult. The traditional view of the nature of ash manna survived into the following century (Weckero [1605], 1617: pp. 365–366; Pemel, 1652–1653: ch. 44, n. p.)

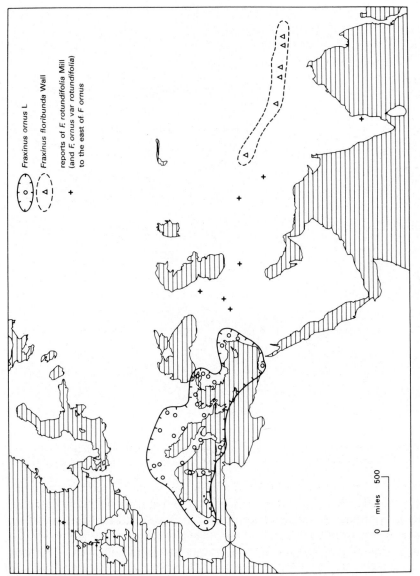

Map 14. Approximate distributions of *Fraxinus ornus* L., *F. floribunda* Wall., and *F. rotundifolia* Mill.

89

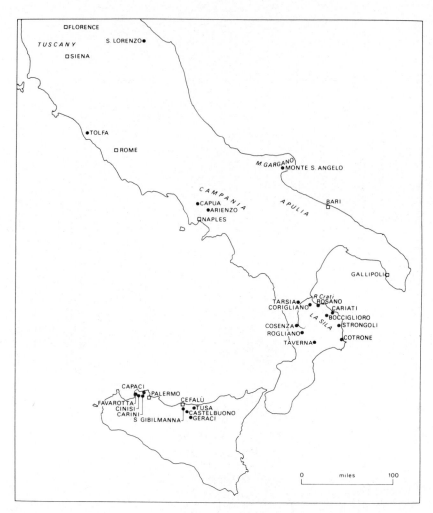

Map 15. Manna of *Fraxinus* spp. in central and southern Italy and Sicily. Places mentioned in 16th to 19th century sources.

the earliest references to artificial incisions,[24] which produced a more copious supply. This, however, was at first thought to be inferior to the "natural" product. Annibale Briganti (ca. 1558) investigated the matter in the producing areas and came to the correct conclusion that there was no essential difference. Briganti's report was published in 1562, without acknowledgement, by the Neapolitan physician Donatus Antonius ab Altomari.[25]

---

[24] Matthiolus, 1558: p. 74. See also *Ricettario Fiorentino* (Colegio de' Medici), 1567: pp. 45–46; 1597: p. 48 ("fatta con arte," "fatta per incisione"). Not mentioned in the first edition, 1498, under *Della Manna* (1968: Primo Libro) or the editions of 1548: pp. 20–21; 1550: pp. 28–29. The *Ricettario* (1567, 1597) refers particularly to the area around Cosenza.

[25] Altomarus, 1562: p. 17; quoted with approval by most subsequent authorities, for example Dale, 1693: p. 496; Boccone, 1697a: p. 79; Geoffroy, 1741: 2: pp. 587–588. Concerning the origin of the observations, see Briganti in Orta, 1576: p. 20.

Garcia da Orta (1563) and Christovão Acosta (1578), in accounts of the pharmacopoeia of the Indies, compared Calabrian manna with the better known Oriental varieties.[26] The former was imported through Venice, according to Leonhard Rauwolf (1573–1576).[27] The "manna of San Lorenzo" was included among the products of the states of the Church in 1589.[28]

It cannot be shown that supplies also came from Sicily until late in the following century.[29] Only Calabria is mentioned in Deusingius's *Dissertationes de manna* (1659). Samuel Bochart (1599–1667) however added Sicily as well as Apulia in his *Geographia Sacra*.[30] He was followed by Pierre Pomet (1694)[31] and by the naturalist Paolo [Silvio] Boccone of Palermo (1697).[32] Boccone clearly had considerable local knowledge and he names a number of places in Calabria and Sicily that produced *manna medicinale*. The reputable manna of Apulia came from the Gargano peninsula,[33] especially the southern slopes around Monte S. Angelo.[34] Tolfa (north west of Rome),[35] Capua in Campania,[36] and the *maremma di Siena*[37] are also mentioned before the close of the 17th century.

(b) Management of Fraxinus spp.

Fraxinus ornus [*rotundifolia*] was conserved for the purpose of obtaining manna from at least the second half of the 16th century. Altomarus (following Briganti) observed that trees were bought and sold. *F. excelsior* also was cultivated, usually in moister situations, before the close of the 18th century.[38] Plantations (*frassinetti*) gradually took the place of natural coppices (*frasseto, bosco di frassini*), more especially in Sicily.[39] "The trees, which attain a height of from 10 to 20 feet, are planted in rows and stand about 7 feet apart, the soil between being at times loosened, kept free from weeds, and enriched by manure. After a tree is 8 years old and when its

---

[26] Orta, 1913: p. 281; Acosta, 1585: pp. 308–310.

[27] Rauwolf (1581), 1693: p. 85.

[28] Ranke, 1908: 1: p. 303. S. Lorenzo I take to be S. Lorenzo in Campo in the March of Ancona.

[29] Fazelli's important book on Sicily (1558) has a section entitled *De Ubertate Siciliae*, but this contains no reference to manna.

[30] Bochart, 1692: 3: col. 873.

[31] Pomet, 1694: 1: p. 236; 1709: p. 33.

[32] Boccone, 1697a: pp. 79, 82 (*manna forzata*); Lémery, 1699: p. 470. Calabria alone in Lovell (1665: p. 490) and Dale (1693: p. 495).

[33] J. Bauhin (1541–1613), 1650–1651: 1: p. 183; Johnstone, 1662: p. 334; Targioni-Tozzetti, 1768–1779: 6: p. 424. Horace (1968: pp. 128–129), who was born (65 B.C.) at Venusia near the border of Apulia and Lucania, referred to the ash trees (*orni*) of the Gargano.

[34] Pomet, 1694: 1: p. 236; 1709: p. 33; Lémery, 1699: p. 470, who also mentions Galliopoli (Gallipoli), south east of Taranto.

[35] Pomet, 1694: 1: p. 236; Boccone, 1697a: pp. 79, 82; Lémery, 1699: p. 470. See also Geoffroy, 1741: 2: p. 594.

[36] Robinson (1683–1684), 1717: p. 474 (probably not harvested).

[37] Boccone, 1697a: p. 79.

[38] Woodville, 1790–1794: 1: p. 105. Cleghorn (1870: p. 133) noted *F. excelsior* around Cefalù but not around Palermo.

[39] Sestini, 1788: p. 92.

stem is at least 3 inches in thickness, the gathering of manna may begin; and may continue for 10 to 12 years when the stem is usually cut down, and a young one brought up from the same root takes its place. The same stump thus has often two or three stems rising from it."[40]

H. Cleghorn (1870) has left perhaps the best description of the propagation and culture of the manna ash, based on observations in northern Sicily:

"*F. ornus* flowers only every third or fourth year, and when it flowers and seeds abundantly it gives no manna . . . . It produces best in low, sunny sites, or the southern slopes of a hill-side . . . . It may be propagated either by seed or by suckers, but the first is much preferred . . . . The ground should be well crumbled and manured. The sowing may take place either in autumn or spring, but the former is preferable. At the beginning of winter of the second year, the young plants should be thinned to an interval of 18 inches. When about 3 feet high, and as thick as a finger, they are transplanted to holes of 18 inches cube, and placed in their permanent site at about 7 feet apart . . . . Frequent weeding is important, and manure should be applied every two years . . . . If the trees are to be introduced into a wood, it is sufficient to make trenches 4 inches deep, drop in the seeds, and cover them. The natural moisture and shade will suffice to maintain the plants."

"Culture is limited to opening up the roots in December, throwing up the soil in March, and levelling again in April . . . . It is important to get the stems to go straight, and to hinder low branching. The trees, however, should not be pruned, but only dead twigs cautiously removed. It is not usual to graft, but if a tree be unproductive, it should be cut down, and a sucker, grafted from a productive tree, tended in its stead. Sometimes the common ash is grafted with the manna ash. Till the stem is of good size, all shoots from the root should be removed, but at a later period they should be cherished, in order to have a substitute ready when the original stem is exhausted."[41]

Several other authorities have described the lopping of exhausted stems and their replacement by a succession of shoots from the stump.[42] Wild trees were apparently managed in a similar but less systematic way.[43]

(c) Collection of manna

The preferred manna of *Fraxinus* spp. was light and friable and pale yellow in colour. The weather most favourable for production, as Cleghorn observed in 1870, "is that in which there are steady north and north west winds, dry air, moderate heats, and calm nights . . . . In wet weather or *sirocco* the manna dissolves and cannot be collected."[44] The desirability of

---

[40] Flückiger and Hanbury, 1879: p. 411.
[41] Cleghorn, 1870: pp. 133–134.
[42] Ward, 1893: p. 381; Hare *et al.*, 1905: p. 952.
[43] Swinburne (1777–1780), 1783–1785: 1: p. 287 (Calabria).
[44] Cleghorn, 1870: p. 134. Cirillo ([1766] 1771: p. 236) referred to the harmful effects of southern winds and of wet weather.

"moderate" temperatures may help to explain the absence of references to ash manna in the Levant and the Near East generally; but this could also be the result of neglect in regions where other mannas were available.

Manna that exuded spontaneously (*spontanea*) was collected in late June, a month or so earlier than that resulting from incisions (*forzata, forzatella*). The 16th- and 17th-century authorities distinguished between manna of the trunk (*di corpo, di ligni, di trunci*) and that found on the leaf (*di foglia*).[45] The latter was highly regarded, but difficult to collect, and could not be artificially stimulated. It had declined in importance by the middle of the 18th century. Geoffroy (1730's) reported that *manna di frondo* was "rarely met with in the shops of Italy";[46] and according to Dominico Cirillo (1766), "those who are employed in the gathering of manna know of none that comes from the leaves."[47]

That the punctures of insects promoted the "spontaneous" flow of manna from the trunk and larger branches was known to countryfolk from the time of Briganti. In 1683–1684, near Capua in Campania, Tancred Robinson "observed a species of ash, or *ornus*, on the trunk whereof many saccharine concretions were visible. This proved [to be] the true *manna*.... Swarms of *cicadas* [*Cicada orni* L.] were sucking the body and boughs, and perhaps by wounding them made way for fresh manna."[48] Robinson was informed by a local physician that "*cicadi* did feed much upon the *ornus* [which] in many places north west of Naples afforded manna...though it was not so much esteemed as that of Calabria."[49]

Transverse incisions were made, daily and in succession, beginning near the base of the trunk, from late July through August and September to October, depending on the onset of the wet season. The notches (about two inches long and one inch apart vertically) were confined to "one side of the tree, the other side being reserved till the year following, when it undergoes the same treatment."[50] In Sicily a peculiar hooked knife was employed. The result of this operation and the work of collection were well illustrated by J. P. Houel in 1782 (Fig. 12).[51]

---

[45] Brasavolus, 1537: p. 335; *Ricettario Fiorentino*, 1548: pp. 20–21; Renodaeus, 1609: p. 274 (*manna de folio*); Pemel, 1652–1653: ch. 44, n.p.

[46] Geoffroy, 1741: 2: p. 593. *Manna di foglia* is mentioned by Savary des Bruslons, 1742: 2: p. 1183; Fothergill, 1746: p. 93; Rolt, 1761: m. p. (*di fronda*, under "Manna": "the leaves being found so loaden with these grains, that they seem covered with snow.")

[47] Cirillo, 1771: p. 235.

[48] T. Robinson, 1717: p. 474.

[49] Robinson in Ray (1685), 1848: p. 176. See also Johnstone, 1662: p. 334; Michaélis, 1774: p. 36; Ritter, 1822–1859: 14: p. 671; Anon., 1828: p. 262; Merat and Lens, 1829–1834: 4: p. 220; Rosenmüller (1830), 1840: p. 320; Leunis, 1844–1853: 1: p. 319; Guibourt, 1849–1851: 2: pp. 533–535; Bach, 1857: p. 289 (*Tettigonia orni*, in southern Germany); Kolb, 1892: pp. 2, 13; Ebert, 1908: p. 428. Juel (1913: pp. 189–195) observed in the botanical garden of Upsala a honeydew (*manna-regn*) voided by *Psyllopsis fraxini* Först. on *Fraxinus excelsior*; the substance consisted of trehalose and saccharose.

[50] Woodville, 1790–1794: 1: p. 105. See also Brydone, 1773: 2: p. 279. Cleghorn (1870: p. 135) described a method of wounding the tree that was apparently peculiar to Tuscany: removing a portion of the bark (about 2 inches by 4 inches) "from the sunny side of the tree. From this the manna continues to exude for about twelve days, and then the wound cicatrises, and a new wound is made. This is done some ten times."

[51] Houel, 1782–1787: 1: p. 53.

Fig. 12. Collecting manna, Sicily (Houel, 1782–1787: 1: pl. 32).

Manna flowed chiefly between noon and early evening. That which ran down the trunk, sometimes to the ground, gathered impurities and was least valued (*grassa*, *rottame*).[52] To prevent this, a receptacle of some kind (commonly the flat *penca* of the opuntia cactus) was sometimes placed near the base of the trunk, or a leaf was inserted in one of the notches. The superior, less glutinous product, usually from the middle notches of young trees and gathered at the height of the season, dried on the bark (*manna en larmes*, "tears") and was carefully removed with a wooden spatula. The highest quality was obtained by placing a piece of straw or a reed in the fresh wound and allowing the exudate to form a projecting "pipe" (*manna canellata*, *manna in cannoli*)[53] (Fig. 13). Manna of mixed quality, including scrapings from the bark, was marketed as *manna en sortes*, "small manna," or *tolfa manna* (named from the town of Tolfa near Civitavecchia).[54] This was only about one-third of the value of *manna canellata*.

Manna was collected in unglazed pots, wooden bowls, or rush baskets (metal was considered harmful), placed on shelves to dry and harden, and then packed in boxes ready for distribution. Cirillo (1766) remarked that there was no need to adulterate the product, as there was normally more available than could be exported.[55] Occasionally, however, there were attempts to market mannas from other regions, within and beyond Italy, as "Calabrian" or "Sicilian,"[56] and inferior grades might be incorrectly described.

Paolo Boccone's *Museo di Fiscia e di Esperienze* (1697) provides the first considerable list of manna-producing townships.[57] Those in Calabria (Cosenza, Cariati, Lucciro, Tarsia, Taverna) lay around the mountains of La Sila (Map 15). J. H. von Riedesel (ca. 1767) referred to Corigliano and Strongoli, as well as to Cariati where "they collect the best manna, and in the greatest quantity."[58] Supplies from the region of Cariati were still "a considerable branch of commerce" in the 1820's,[59] but fifty years later David Hanbury found little or no activity anywhere in Calabria.[60] In Sicily commercial production appears to have been largely confined to two areas – the neighbourhood of Cefalù and the coastal lands to the west of Palermo. Boccone mentioned Capaci, Carini and Castelbueno, Domenico Sestini

---

[52] Pemel, 1652–1653: ch. 44, n. p.; Cirillo (1766), 1771: p. 235; Don, 1831–1838: 4: pp. 56–57; Cleghorn, 1870: p. 134; Ward, 1893: p. 381; Wood and Bache, 1907: p. 765; Hare *et al.*, 1909: p. 952.

[53] Geoffroy, 1741: 2: p. 593; Fothergill, 1746: p. 94; Cirillo (1766), 1771: p. 235; Sestini, 1788: p. 97 (*in cannuola*); Green, 1820: 1: p. 580; Don, 1831–1838: 4: pp. 56–57; Cleghorn, 1870: p. 134. Cf. Ward, 1893: p. 381.

[54] Alibert, 1814: 1: pp. 313–315 (ranked after the manna of Calabria, the Gargano and Sicily); Janssen, 1879: p. 407; Flückiger and Hanbury, 1879: p. 412; Planchon and Collin, 1895–1896: 1: p. 745; Wood and Bache, 1907: p. 764; Hare *et al.*, 1909: p. 952. Also known as *manna Capaci* and *manna Geraci*, two places in Sicily (Jourdan, 1828: 2: p. 10).

[55] Cirillo, 1771: p. 236. Cf. More, 1750: p. 471 (informed of "ways of counterfeiting the several appearances of [manna]").

[56] Robinson (1683–1684), 1717: p. 475. Cf. *Ricettario Fiorentino*, 1548: p. 21; J. Bauhin (1541–1613), 1650–1651: 1: p. 197.

[57] Boccone, 1697a: p. 79.

[58] Riedesel, 1773: p. 165.

[59] Craven, 1821: pp. 224–225.

[60] Hanbury (1872), 1876: pp. 365–367 (Cosenza, Corigliano, Rossano, Cotrone, Rogliano). Cf. Janssen, 1879: p. 407.

De la Manne.

Manne de Briançon

Manne Liquide

Manne de Calabre

Fig. 13. Manna-producing species (Pomet, 1694: 1: p. 236): *Fraxinus ornus* L. (Manne de Calabre), *Alhagi* sp. (Manne Liquide), *Larix europaea* D. C. (Manne de Briançon).

(1788) Tusa and Cefalù itself.[61] In 1870 the ash plantations of northern Sicily covered about 8000 acres.[62] Each hectare (2.5 acres) of 4000 to 5000 trees might be expected to yield about 2000 pounds of manna.[63]

There was no serious competition from elsewhere in Italy, except perhaps in the 16th and 17th centuries from parts of Apulia, including the Gargano peninsula. Neither Tuscany[64] nor Campania was of major importance. Robert More (ca. 1750) wrote of the collection of manna at Arienzo, between Naples and Benevento.[65] Cirillo (1776), on the other hand, observed that the ash trees of this area were unproductive "for want of cultivation."[66]

The manna of *Fraxinus* spp. contains 60 to 80 per cent mannite.[67] It was used exclusively in medicine. Demand throughout Europe increased from the 16th to about the middle of the 19th century.[68] By the second half of the 18th century, its reputation generally surpassed that of the various Oriental products.[69] Market prices varied with supply and this, in turn, largely according to weather conditions in Calabria and Sicily. Manna was one of the few readily exportable products of Calabria in particular, and for 300 years or so it held an important place in the regional economy. "The King of Naples," More maintained, "has so large a revenue from [manna] that he is extremely jealous of it, [and] during the season guards the woods by *sbirri*, who even fire upon people that come into them, and he makes the stealing of the liquid, death."[70] According to Riedesel, "the owners of the trees [were] obliged to sell their manna to the King for a fixed price, the better sort, or what is commonly called *in canole*, for two *carlini*, and the worse, or *in frasca*, for eight *grani* the pound. These revenues [were] farmed for 32,000 ducats per annum"[71] Other eye-witness accounts suggest that the "contractor" system was widely abused and led to hardship and to discontent among the peasant population.[72]

---

[61] Sestini, 1788: p. 93. For these and other places see also Hanbury (1872), 1876: p. 367; Flückiger and Hanbury, 1879: p. 411; Ward, 1893: p. 381; Planchon and Collin, 1895–1896: 1: p. 745.

[62] Cleghorn, 1870: p. 137.

[63] Simmonds, 1895: p. 134.

[64] Boccone, 1697a: p. 79; Cleghorn, 1870: p. 135; Hanbury (1872), 1876: p. 365; Flückiger and Hanbury, 1879: p. 410.

[65] More, 1750: pp. 470–471.

[66] Cirillo, 1771: p. 234.

[67] Soubeiran, 1840: 1: pp. 639–641; Stillé, 1868: 2: pp. 437–439; Royle (1847), 1876: p. 524; Flückiger and Hanbury, 1879: p. 412; Flückiger, 1883; pp. 22 ff; Hare *et al.*, 1905: p. 953; Wood and Bache, 1907: p. 765; Wehmer, 1929: pp. 596–597.

[68] Cleghorn (1870: p. 135) reported that "merchants concur in saying that demand is falling off." He gives the value of manna exported from Sicily in 1852 and in 1863–1866 to the United States, the Baltic, France, Great Britain and the colonies, Italy, and "other countries." Cf. Ward, 1893: p. 381 (100 tons annually to England ca. 1870); Bentley and Tremen, 1880: 3: nos. 170–171 (export of 350 tons from Sicily in 1872); Simmonds, 1895: p. 135 (about 200 tons from Italy, 1884).

[69] Rolt (1756), 1761: n.p. (under "Manna").

[70] More, 1750: pp. 470–471.

[71] Riedesel (ca. 1767), 1773: p. 165 (one *carlino* = ca. 4 pence). Cf. Brydone, 1773: 2: p. 279 ("of all the variety [of trees] that is cultivated in Sicily, the manna tree is esteemed the most profitable.") Fiore da Cropani (1691–1743: 1: p. 253) reported that 30,000 pounds of manna per annum from Campania and Boccigliero (Calabria) were worth 1100 ducats in excise (*gabella*). Manna is mentioned in the *Tariffa delle Gabelle per Firenze*, 1791: p. 57.

[72] Swinburne (1777–1778), 1783–1785: 1: pp. 288–289.

C. GLYCERIA sp.

The term "manna" has been applied to several small, edible seeds, more particularly to the fruit of the wild *Glyceria fluitans* R. Br., the *Festuca fluitans* or "gramen aquaticum fluitans" of Linnaeus (1753).[73] This was formerly collected on a considerable scale in parts of Poland, Brandenburg, Pomerania, East Prussia, the central lowlands of Hungary, and probably to a lesser extent in Bohemia, southern Sweden, Denmark, and White Russia (Map 16).[74] The eastern limits of use have not been satisfactorily determined. The most important area in the early modern period (16th to 18th centuries) appears to have extended from the middle Oder (giving rise to the descriptions "Gramen Mannae Francofurtanum" and "Frankfurter Schwaden") eastward to Poznan and Warszawa. The ethnic association is predominantly Slav and probably goes back to at least the early medieval period when agriculture was less important. *G. fluitans* was perhaps first noticed by Albertus Magnus (ca. 1193–1280) in his *De Vegetabilibus*.[75] "Schwaden" (swath), "Schwadengras" and "Manna Schwaden" were the usual descriptions in lands settled or re-settled by the Germans. We also find "Mannagras," "Mannaschwingel" (*schwingel*, fescue), "Mannahirse" (*hirse*, millet), "Mannagrütze" (*grütze*, groats or grits), "Schwedengrütze" and, in Hungary, "mannakása" (*kása*, groats; *kásafú*, wild millet grass). The 16th and 17th-century herbalists (Figs 14, 15) refer to "Gramen Mannae," "Manna Graminea," "Manna Polonica" (*Mannapolska*) and "Manna Germanica."[76] Ladislaus Bruz (1775) added "Manna Borussica" (Belorussiya), "Manna Hungarica," "Manna Francofurta" and "Manna Prutenica" (Prussia).[77]

G. fluitans is sometimes said to have been cultivated,[78] but such statements seem to involve confusion with the cultivated *Panicum* [*Digitaria*]

---

[73] Linnaeus, 1957–1959: 1: p. 75, no. 10. Less commonly, *G. plicata* Fries, *Panicularia fluitans* O. Kuntze, *Poa fluitans* Scop., *Hydrochloa fluitans* Hartm., *Molina fluitans* Hartm. (Ascherson and Graebner, 1898–1902: 2, 1: p. 446). Gast (1968: pp. 243–244) reported that "sweet manna" is found in the Ahaggar of southern Algeria on the Gramineae *Imperata cylindrica* Beauv. [*Arundinacea cyrilli*] and *Erianthus ravennae* Beauv. James Bruce ([1768–1773] 1790: 5: pp. 47–48), writing of Abyssinia, observed that "on the leaves of some [wild Gramineae] I have seen a very small glutinous juice .... this is of the taste of sugar."

[74] Bruz, 1775: p. 26; Ascherson, 1864: p. 850; 1895–1896: pp. 37–60; Hackel, 1887: p. 74; Ascherson and Graebner, 1898–1902: 2, 1: pp. 446–447; Hartwich and Håkanson, 1905: pp. 473–478; Fedtschenko, 1928: pp. 191–192; Komarov, 1934–1962: 2: p. 451. The plant itself is found much more widely and is relished by horses, cattle and geese (thus Danish *gås gras*): in temperate Europe (including Great Britain, and up to 1740 metres in the Alps), Asia (including Siberia and Japan), North West Africa, Australia, and North and South America (Schumacher, 1801: 1: p. 31; Steven, 1811: p. 9; 1856–1857: 30, 2: p. 109; Berchtold, 1836–1842: 1: pp. 250–251; Johnson, 1862: p. 284; Lindley and Moore, 1874: 1: p. 536; Hermann, 1956; pp. 133–134). G. Usher (1974) states that the seeds were eaten by the Indians of North America. Casual collection may have been practised in many areas.

[75] Albertus Magnus, 1867: p. 633. Cf. Konrad von Megenberg (ca. 1309–1374), 1897: pp. 72–73.

[76] Matthiolus, 1565: p. 1000; Dodonaeus, 1583: p. 549 ("Gramen Mannae primum" and "Gramen Mannae alterum"); Dalechamps, 1586–1587: 1: p. 414; Gerarde, 1597: p. 25; Johnstone, 1662; p. 338; Lovell, 1665: p. 183; Salmasius, 1689: 2: p. 254. Including both *Glyceria fluitans* and *Panicum sanguinale*.

[77] Bruz, 1775: p. 11.

[78] Syme, 1872: p. 97 ("in several parts of Germany").

Map 16. Approximate areas of collection of "Polish manna" (*Glyceria fluitans* R. Br.); 16th to 19th century sources.

sanguinale L. ("Bluthirse") and possibly *P. italicum* L.[79] (*Setaria italica* Beauv.). The name "manna" was, however, sometimes applied to *Panicum* spp., and might also refer to the flour or meal of various grains, alone or mixed (Polish, *kasza manna*; Russian, *mánnaya krupá*).[80] Typically, the seeds of *G. fluitans* were crushed in a wooden mortar and made into a kind of gruel.[81] Slightly sweet ("sugar grass"), it was often preferred to panic millet (*Panicum miliaceum* L.), and until about the middle of the 19th century the product was traded and exported to various parts of northern Europe, chiefly through Danzig and Königsberg.[82] John Gerarde (1597) observed that it was "sent into Middleborough and other townes of the

---

[79] Hartwich and Håkanson, 1905: pp. 473–478. Maurizio ([1916–1917] 1932: pp. 77, 80–81) discusses this question in relation to the works of, among others, P. Crescentius (1571) and D. S. Sirrenius (1613).

[80] Baxter and Johnson (1934) give the date 1461 for manna, "flour."

[81] Jourdan refers to *G. fluitans* in his *Pharmacopée universelle* (1828: 1: p. 551) but its medicinal use appears to have been very limited.

[82] Krünitz, 1808–1828: 149: pp. 722–728.

Fig. 14. *Glyceria fluitans* R. Br. (Matthiolus, 1565: p. 1000).

Lowe Countries in great quantitie . . . ."[83] In the 17th century the seed was included among rents in kind, at least in Poland. *G. fluitans* was probably unsuitable for cultivation for it thrives in very wet or marshy situations (thus "Wasserschwaden," "fluthende Schwaden," "floating sweet grass," "flote fescue," "floating glyceria"), and its decline as a food grain can be broadly

---

[83] Gerarde, 1597: p. 25. "Small quantities" were imported into London in the second half of the 19th century (Johnson, 1862: p. 284). John Smith (1882: p. 265) states that *G. fluitans* was consumed in Holland (and in Poland and Germany); but whether this was imported or collected locally is not clear.

GRAMEN MANNAE, CAP. XXI.

Li.17.ca.8.

Il ti generibus quidam, alij Panici, adſcribunt plantam quam Germani vocant Gramen Mannæ. Huius duæ ſunt ſpecies. Alteram priuatim capriolam, & Sanguinellā nominant, quidam Iſchæmon Plinij eſſe volunt, de quo poſtea nos agemus. Gignitur ſponte in pleriſque Italiæ, Germaniæ, Boëmiæ locis incultis. Goritienſes tamen & Carinthij ſerunt. Fibroſa radix eſt, craſſa, in tranſuerſum porrecta, ſtipulæ cubitales, ac nonnunquam proceriores : crebro geniculatæ, cum maturuerint, ſubrubentes : Arundinacea ſunt, aut Gramini ſimilia, & villoſa folia, præſertim quæ culmos ambiunt, & amplectuntur. Iuba, vt Milij, effuſa, minus tamen denſa, nigricās, in ſpicas diſſecta, longas, & tenues, ſemen ferentes ab vna tantū parte, Milio minus, oblōgum, quod in pilis à glumarum & vtriculorū inuolucro purgatum, candidū eſt Oryzæ modo. Eo Bohemi cocto cum iure pinguis carnis veſcūtur, & ſic paratum, ſuaue gratúmque edulium eſſe prædicant. Sclaui vocant Cornicis pedé, qua nomenclatura deceptus Leonicemus Coronopū Dioſcoridis eſſe credidit. Alterum Mannæ gramen in agrorum aggeribus, Germanię, Belgij, Galliæ, cæterarúmque regionum Europæ, nullo ſatu gignitur: quidam tamen & in hortis colunt. Arundinacea huic quoque natura eſt. Radix multas fibras ſpargit. Arun-

GRAMEN MANNAE
prius Matthioli.

GRAMEN MANNAE
prius Dodonæi.

GRAMEN MANNAE
secundum Dodonæi.

dinifo

Fig. 15. *Glyceria fluitans* R. Br. (Dalechamps, 1586–1587: 1: p. 414).

101

associated with the large-scale reclamation of land for cultivation. By the early 20th century it had all but been forgotten. "En 1914," wrote D. A. Maurizio, "M. B. Issatschenko m'informa qu'on pouvait se procurer cette farine dans les villes russes, mais seulement comme aliment de fantaisie. On ne le connaît pas d'une façon générale."[84]

The manna-grain was harvested between May and July by striking the rachis of the plant, thus shattering the ripe ear; it was then dried in the sun.[85] Teodor Zawacki (1616) mentions "manne trząść" (*trząść*, to shake).[86] E. Connor (1690) observed that "In [the] Palatinate [of Cracow] and some others, there is a particular sort of manna, which they gather in the months of May and June by sweeping it off the grass with sives . . . . They eat this manna, and make several sorts of dishes with it . . . ."[87] This appears to refer to the collection of *G. fluitans*. Linnaeus has the following description: "The seeds of [*Festuca fluitans*] are gathered yearly in Poland, and from thence carried into Germany, and sometimes into Sweden, and sold under the name of manna-seeds (*seminum mannae*). These are much used at the tables of the great, on account of their nourishing quality and agreeable taste. It is remarkable that amongst us these seeds have hitherto been neglected, since they are so easily collected and cleansed."[88] The earliest known work devoted to the subject is by J. S. Ledel: *Succincta Mannae Excorticatio Betrachtung des Schwadens* (Sorau, 1733). This was followed by S.-M. Hillscher's *Prolusio de Gramine Manna Dicta* (Jena, 1747) and by Ladislaus Bruz's more useful and better known *Dissertatio inauguralis de gramine mannae, sive festuca fluitante* (Vienna, 1775).

The appellation "manna" is both interesting and obscure. Like most other mannas, the grain of *G. fluitans* was a natural product (*mann*, in the sense of "gift") with a somewhat sweet taste, and locally it was much appreciated – even something of a luxury and finally no more than a traditional speciality.[89] Although intrinsically very different from the others, it is usually considered, or at least mentioned, in early works on manna and *materia medica*. Somehow the ancient and persistent notion of manna as a "dew," *ros coeli*, descending at night was applied, in whole or in part, to *G. fluitans*. Thus we have "Himmelstau" and "harmat-kása" (Hungarian); also "Hexentau" (witches' dew) and "boszorkány-kása" (magic groats).[90] According to Connor, the manna was collected "together with the dew" (? in the early morning). A. Stillé observed that the description *ros meleus* (honey dew) "is also given to a substance resembling millet seed, which is said to have fallen from the air in the confines of Silesia and Poland and in other

---

[84] Maurizio (1916–1917), 1932: p. 78.

[85] Krünitz, 1808–1828: 149: pp. 722–728; Ascherson, 1895–1896: p. 43.

[86] Zawacki, 1891: p. 40, no. 246; see also p. 58, no. 397.

[87] Connor, 1698: 1: p. 248. Repeated by Savary des Bruslons (1723) 1751–1755: 2: p. 484 (without acknowledgement). Bruz (1775: pp. 29–34) states that the grain was collected in Hungary in May, in Poland in June and July.

[88] Linnaeus, 1786: 3: p. 80, no. 90; translated by Stillingfleet, 1762: p. 386. Cf. Rousseau, 1794: p. 139.

[89] Willemet (1808: 1: p. 94) claimed that in Prussia the grain was used for making beer as well as bread "en temps de disette."

[90] Bruz, 1775: p. 20; Hartwich and Håkanson, 1905: p. 474.

places, and which was used as food."[91] Earlier, Benjamin Stillingfleet put forward a more persuasive explanation: "There is", he wrote, "a clamminess on the ear of the *flote fescue*, when the seeds are ripe, that tastes like honey . . . . and for this reason, perhaps, they are called manna-seeds."[92]

D. LARIX sp.

The Paris Customs Tariff of 1542 is said to include *manna brianzona* or *brigantianca*,[93] found on the leaves of the European larch, *Larix europaea* D.C. (*Pinus larix* L.), at high elevations around Briançon, Dauphine.[94] Belon du Mons (1550) described the product,[95] and Lobel (1570–1571) published an illustration of the larch (Fig. 16).[96] Rauwolf (1581) compared the "manna which we gather from the *Larix*" with *trunschibel*,[97] that is *tar-angubīn* from *Alhagi maurorum* and *A. camelorum*.

Several 17th- and 18th-century authors refer to *manna laricea*,[98] from which we gather that it was available in only small quantities and that its medicinal reputation was second to that of the ash manna of Calabria and Sicily. The best description is by Dominique Villars (1788).[99] Larch manna was collected, usually by local shepherds, in the early morning during the hottest and driest part of the year (June and July). Young trees or recent growth on older specimens provided the chief supply. The product was costly and apparently not much used in the early decades of the 19th century.[100] Its chemical composition was first established by Marcellin Berthelot (1859) who named the principal constituent, a peculiar sugar, *mélèzitose* (French *mélèze*, larch).[101] A few years later, David Hanbury

---

[91] Stillé (1860), 1868: 2: p. 438. The comment may come from Pomet (1694: 1: p. 234), quoting the Jesuit Cornelius a Lapide (died 1637). See Lapide, 1866: p. 582 (". . . . manna Polonicum, quod in Polonia (Polonis omnibus attestantibus) mense junio et julio noctu depluit, herbisque instar roris incumbit."). Cf. Dodonaeus, 1583: p. 549; G. Bauhin, 1671: p. 497 ("Manna Germanica liquida flava . . . .")

[92] Stillingfleet, 1762: p. 386. Lovell (1665: p. 183) has "dew grass, *gramen mannae esculentum.*"

[93] Henry, 1924: p. 387.

[94] Specifically, on "the slopes on the right bank of the Guisanne, on the left bank of the Cerveyrette, on the right bank of the Durance, and in the valley of the Guil in the Queyras" (Henry, 1924: p. 387, based on local information).

[95] Belon du Mons, 1553: p. 9a (*briansona*); 1555: p. 129.

[96] Lobel, 1576: pp. 24, 449; 1591: 1: p. 50.

[97] Rauwolf, 1693: p. 84.

[98] Renodaeus, 1609: pp. 274–275 (*manna larigna, manna briansonnensis*); J. Bauhin, 1650–1651: pp. 183 (*manna laricis*), 191 (*manna laricea sive briansona*); Johnstone, 1662: p. 334; G. Bauhin, 1671: p. 497 (under *manna officinarum*); Pomet, 1694: 1: pp. 238–239; 1709: p. 33; Alexandre, 1716: p. 370; T. Robinson 1717: p. 475; Labat, 1730: 5: p. 314; Geoffroy, 1741: 2: p. 590; Duhamel du Monceau, 1758: 1: p. 151; I. E. Fabri, 1776: p. 102; Sestini, 1788: p. 92; J. A. Murray, 1793: 1: p. 21 (*manna laricea, brigantina*).

[99] Villars, 1786–1789: 4: pp. 808–809.

[100] Krünitz, 1808–1828: 83: p. 737; Loiseleur-Deslongchamps, 1819: 2: p. 521; Moringlane *et al.*, 1822: p. 335 ("24 frs. d'une once"). It is, however, included among *materia medica* in Merat and Lens, 1829–1834: 4: p. 219, and Guibourt, 1849–1851: 2: pp. 241, 534.

[101] Berthelot, 1859a: pp. 282–286; 1859b: pp. 61–64. Cf. Bonastre, 1833: pp. 443–447; Alëkhine, 1889: pp. 532–551.

MANNA è RAMIS LARICIS.

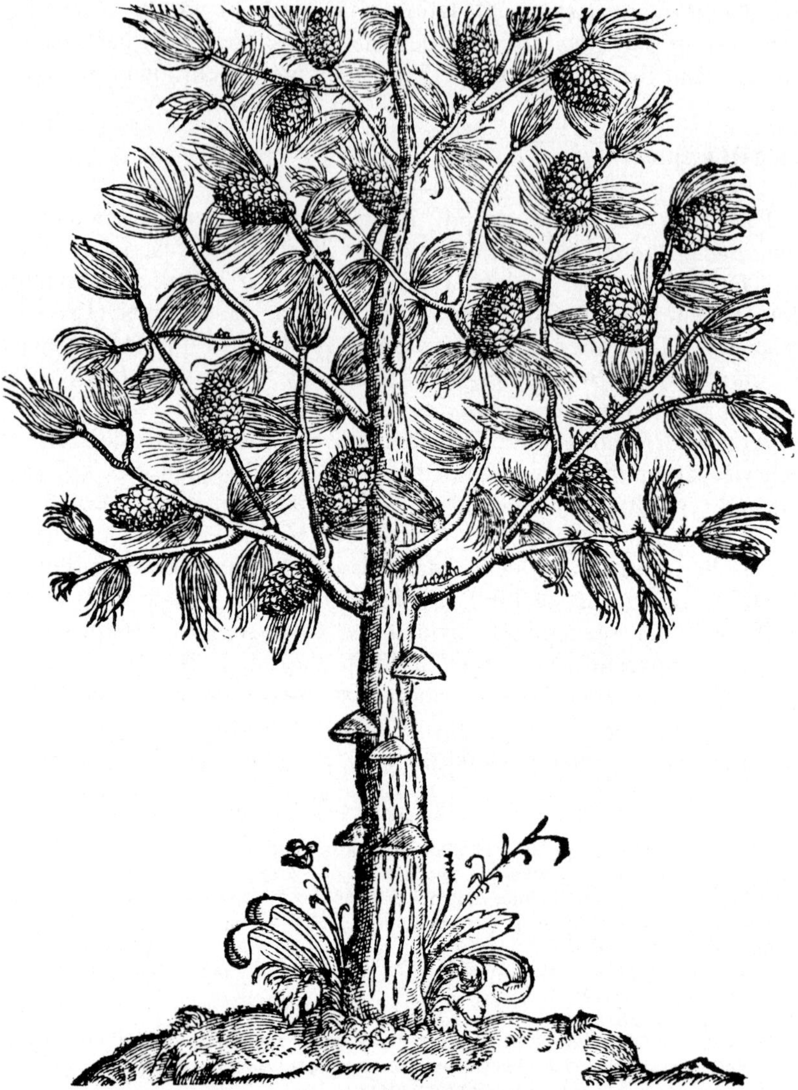

Fig. 16. *Larix europaea* D. C. [*Pinus larix* L.] (Lobel [1570–1571], 1576: pt. 2: p. 24).

managed to obtain a sample of the manna from Chantemerle near Briançon.[102]

The majority of reports of larch manna refer only to the country around Briançon, but it is probably present in other parts of the highlands of central Europe. John Johnstone (1662) mentioned Styria (eastern Austria).[103] And in 1919 the product was discovered at 4000 to 6000 feet elevation in the

[102] Hanbury (1864), 1876: p. 438. This apparently found its way to Kew (Henry, 1924: p. 387). See also Flückiger and Hanbury, 1879: p. 416; Flückiger, 1883: p. 28.
[103] Johnstone, 1662: p. 334. Cf. Lobel (1570–1571) 1576: p. 449.

Valais of Switzerland.[104] Expert examination suggested that it was not an exudation from the leaves of the larch, as had hitherto been generally assumed, but a honeydew excreted by the aphid *Lachnus laricis*, a parasite of *L. europaea.*[105] In hot and dry summers the honeydew is likely to harden on the leaves of the larch, and thereby be noticed, rather than fall as a liquid. This would help to explain the sporadic occurrence and territorial distribution of larch manna.

E. TILIA sp.

Near Strasbourg, during the hot summer of 1842, Langlois found a saccharine deposit on the upper leaves of the lime or linden, *Tilia europaea* L. (*T. cordata, vulgaris, platyphyllos*).[106] The same substance was noted by Boussingault at Liebfrauenberg in July, 1869.[107]

---

[104] Henry, 1924: pp. 387–388. Simmonds (1895: p. 134) states that it was "eaten in Russia."
[105] Bonastre (1833: p. 477) had suggested that punctures by insects stimulated the flow of manna.
[106] Langlois, 1843a: pp. 444–447; 1843b: pp. 348–351 (including "mannite"). Henry (1924: pp. 387, 390) described this as "honeydew," containing melezitose.
[107] Boussingault, 1872: pp. 214–218 ("c'était une pluie de manne"; insects were not observed). See also Ludwig, 1870: p. 44 ("Linden," quoting Treviranus).

5 MANNAS OF EAST AFRICA AND OF MADAGASCAR

A. EAST AFRICA

When crossing the plateau that separates Lakes Tanganyika and Nyasa (Zambia and Tanzania), A. T. Swann (1883) observed on the ground a substance with "all the characteristics of manna....white in colour, like hoarfrost, sweet to the taste, [and] melt [ing] in the sun....The natives were not allowed to gather it before asking permission of the chief...."[1] Apparently there was no local explanation of its origin. It can hardly have been the "manna" reported (1913) from north west Rhodesia and described as an encrustation on the leaves and twigs of a species of *Gymnosporia* (? *G. deflexa* Sprague).[2]

Two early and curious references to the production of manna are to be found in the works of Pedro Teixeira (*Travels*, 1586–1605) and Friar Joaõ Dos Sanctos (1597). They relate to the Ilhas Kerimbas, a coral archipelago off the coast of Mozambique, south from Cabo Delgado, and including the very small Ilha Teixeira. According to Teixeira, "On the African coast of the Indian Sea, near Mozambique, there are two islands called Aniza [Amiza, alternatively Wamasi or Vamizi, among the most northerly of the group] and Querina [Querimba] wherein much mana (*sic*) is obtained, but of comparatively the lowest quality."[3] The report of Dos Sanctos (published by Samuel Purchas) runs as follows: "In the wildernesses of the Ile Cabo de gado is a store of Manna, procreated of the deaw of Heaven falling on certaine trees, on which it hardens as it were Sugar candide, sticking to the wood like Rozin, and hanging on the leaves, gathered and sold in jarres by the Inhabitants. It tastes sweet as Sugar, in India they use to purge with it. I have often been in the place, and gathered it with my hand. It growes only on one kind, although there be many other trees in the Iland."[4] No more recent account has been found.[5] It is possible that the manna was the same as that collected in western Madagascar from at least the 17th century.[6]

---

[1] Swann, 1910: pp. 118–119. The author claimed that other Europeans could "vouch for the accuracy of [the] description of this food." He suggested that "it might be a mushroom spawn." Discussed by Holmes, 1920: pp. 176–178.
[2] Anon., 1913: pp. 332–333. Furlong and Campbell (1913: p. 128) showed that this substance contained dulcitol (54 per cent), sucrose (6.6 per cent), and dextrose (6.4 per cent).
[3] Teixeira, 1902: p. 204. Repeated by Stevens, 1715: p. 30.
[4] Dos Sanctos, 1905: p. 250.
[5] According to W. F. W. Owen (1833: 2: p. 14), "only Ibo and Querimba [were] inhabited....their commerce consisting solely in slaves."
[6] Pomet (1694: 1: p. 239) and Savary des Bruslons ([1723] 1742: 2: p. 1184) refer to "the manna of Africa," but provide no details.

B. MADAGASCAR

Étienne de Flacourt (1607–1660), who resided in Madagascar for several years, observed in his major *Histoire* of the island (1658): "Il y a une espece de sucre qui est formé par certains papillons.... sur les fueilles d'un arbrisseau, il est dur et doux comme le sucre, les habitans en sont friands, et disent qu'il est tres-souuerain pour la toux et les fluxions sur la poitrine" (pneumonia).[7] This product was known as "*tantelle* [honey[8]] *sacondre* du nom du papilon qui le forme." About 1847, M. Choron of the Collège de Saint-Denis, Ile de Bourbon, submitted a brief report (apparently unaware of Flacourt's comments) on "une substance nouvelle de Madagascar.... [qui] ressemble à de la manne; on la trouve sur certains arbres; les naturels la mangent, et disent qu'elle est déposée par une mouche.... [elle] est considérée comme une gomme dure dans le pays de *Mahafali*, qu'elle s'appelle *taïpinder*, et provient d'un arbre appelé *maroua*."[9] The manna, containing dulcitol, was identified in 1930 as the secretion of the insect *Phremia rubra* Signoret, living on bushes of the family Combretaceae.[10] *P. rubra* is known from most parts of Madagascar; the manna, on the other hand, is found only in the drier west and south west (Màhafàly). It was harvested in September to November and consumed by the Sàkalàva[11] who called the product *tankely sakondy*, "miel de cicadelle" (Cicadellidae or "leaf hoppers"). Commercial production (by expanding the distribution of the Combretaceae) was proposed, but did not prove to be practicable.

---

[7] Flacourt, 1658a: p. 145 (in a discussion devoted largely to *tabaxir*, and quoting Paludanus).
[8] Flacourt, 1658b (*Dictionnaire de la langue de Madagascar*), *tantele, tentele*; 1905, *tintely* (modern); R. Drury (1729), 1890: p. 326, *tentala, tantely*.
[9] Choron, 1847: p. 397. Planchon and Collin in their *Les drogues simples d'origine végétale* (1895: 1: p. 750) have the following note: "La dulcine ou manne de terre, dont l'origine botanique est encore indéterminée. Elle se présente en morceaux grisâtres, souillés de terre, à saveur légèrement sucrée; elle nous arrive de Madagascar et contient de la dulcite, principe analogue à la mannite."
[10] S., F., 1938: p. 54.
[11] The Sàkalàva, comprising several tribes, belonged to the semi-arid South West, but expanded from the 17th century to dominate the western half of the island.

6 MANNAS OF AUSTRALASIA

The aboriginal inhabitants of Australia and Tasmania obtained sweet substances from a variety of sources, chiefly honey, insect honeydew, nectar,[1] the sugar ant *Melophorus inflatus*,[2] and several kinds of "manna." The latter include an insect excretion (*lerp*) and exudations from species of *Andropogon* (*Dichanthium*), *Myoporum* and, most important, *Eucalyptus*.[3] Only the products of *Andropogon* sp. and *Myoporum* sp. contain mannite.

A. ANDROPOGON sp.

In 1897 R. T. Baker and H. G. Smith reported "manna [consisting of 58 per cent mannite] in the form of nodules at the nodes of the stems of the "blue grass" *Andropogon annulatus*" (*Dichanthium annulatum* [Forsk] Stapf) from a location near Townsville, Queensland.[4] It was said that a bushel could be collected in an hour "almost anywhere on the plains," but no direct evidence of the use of this manna as food or in medicine has been found.

B. MYOPORUM sp

Myoporum platycarpum R. Br. ("sandal tree," "sugar tree")[5] is widely distributed in western New South Wales, South Australia and southern Western Australia. The earliest known report of it as a source of resin and of a very sweet manna was published by K. H. Bennett in 1883.[6] During the hottest months of the year, the manna (85 to 90 per cent mannite)[7] exudes from and solidifies upon the trunk and branches of the tree in substantial

---

[1] E. Palmer, 1884: pp. 95 (*Bauhinia carronii* F. v. Muell), 106 (*Banksia marginata* R. Br.); Macpherson, 1939: p. 180; Johnston and Cleland, 1942: p. 99 (*Leptospermum coriaceum* Cheel); Irving, 1957: p. 122 (*Banksia dentala* [? *B. dentata* L.], *B. marginata, Bauhinia carronii, Eremophila* sp., *Leptospermum coriaceum, Telopia speciosissima* R. Br., *Grevillea kennedyana* F. v. Muell, *Hakea lorea* R. Br., *Lambertia formosa* Sm.); Lawrence, 1968: pp. 54, 84, 209, 214.
[2] Irving, 1957: p. 122; Brothwell, 1969: p. 68.
[3] A saccharine exudation from *Leptospermum scoparium* Först. ("tea tree," *manuka*) has also been recorded ([1885] Maiden, 1920: p. 102), but whether this is put to use is not known.
[4] Baker and Smith, 1897: pp. 291–308. References in Hare *et al.* (1905: p. 953), Wood and Bache (1907: p. 764) and Wehmer (1929: p. 44) are based on Baker and Smith. The latter quote the opinion of the entomologist W. W. Frogatt that the manna was perhaps "due to the action of a homopterous insect on the stem of the grass." *A. annulatus* is found elsewhere in Australia and also in Asia and Africa (Forskål, 1775a: p. 173, *Andropogon annulatum*).
[5] Brown, 1810: p. 515.
[6] K. H. Bennett, 1883: pp. 349–351.
[7] Maiden, 1892: pp. 1, 4; Flückiger, 1894: pp. 311–314.

quantities (up to 11 pounds).[8] Whether insects play any part in its production is still obscure. Apparently artificial incisions do not induce "bleeding."[9]

Bennett found that "the natives [of western New South Wales] are very fond of [the manna] and either eat it or, by dissolving it in water, make a kind of drink."[10] On the other hand, around the Fraser Range, Western Australia, where the substance was known as *wairu* or *bulgar yumbu* ("bulgar," *M. platycarpum*),[11] the natives were not partial to it (on account of its laxative properties ?), preferring the gum of *Acacia leiophylla* Benth. and the manna found on *Eucalyptus* spp., including *lerp*.[12]

C. EUCALYPTUS spp.

(a) Lerp

Lerp (*laap, laarp, leup, lárap, layurp*), an insect excretion found on the leaves of several species of *Eucalyptus*, was formerly of considerable importance as an item of food among the natives of Victoria, New South Wales, South Australia, and possibly Tasmania. It was apparently first observed in the late 1830's or the mid 1840's. According to William Westgarth (1848), "Mr Robinson, the Chief Protector, ascertained during his expedition in 1845 to the north west of Australia Felix, that the natives of Wimmera [northern Victoria] prepare a luscious drink from the *laap*, a sweet exudation from the leaf of the mallee (*E. dumosa*). This liquor is manufactured in the months of February and March, on which occasions there is commonly a festival and the adjusting of mutual disputes."[13] Benjamin Bynoe (ca. 1840) probably had *lerp* in mind in referring to a "concreted white substance" the product (*per anum*) of Cicadae during the warmest part of the year. "The natives gather it in their rush baskets and use it as part of their food."[14] Likewise, "a secretion formed on the leaves of the Eucalypti in New Holland by minute *Psylla* .... for several months during the past year [1845–1846], the *womela* had formed a great portion of the food of the natives of New South Wales."[15] P. Beveridge (1884, writing of the Lower Murray, Lower Murrumbidgee, Lower Lachlan, and Lower Darling) found that the search for *laarp* extended up to 20 miles from waterholes; the harvest lasted from six to eight weeks and 40 to 50 pounds could be

---

[8] Chisholm, 1958: 5: p. 479.
[9] Fisher, 1945: p. 164.
[10] K. H. Bennett, 1883: p. 351.
[11] Helms (1891), 1896: pp. 323, 325; Maiden, 1892: p. 1; Cleland and Johnston, 1937–1938: p. 33.
[12] According to Johnston and Cleland (1942: p. 101), the "gum" of *M. platycarpum* is used as an adhesive in the Ooldea region of South Australia. The "manna" is mentioned by Dragendorff, 1898: p. 619; Wood and Bache, 1907: p. 764; Wehmer, 1929: pp. 710, 819.
[13] Westgarth, 1848: p. 73.
[14] Stokes, 1846: 2: p. 483.
[15] Westwood, 1846: p. 659. Cf. G. Bennett, 1860: p. 272 n.

collected in a day.[16] Among the Wirrung of the Ooldea region, South Australia, *lerp* was known as *woma* ("sweet").[17]

The first extended description and analysis of *lerp* appeared in a paper by T. Anderson (1849, based on reports and samples from north and north west of Melbourne).[18] Anderson noted the aggregated "conical cups" of *lerp*, determined that the sugar (49.06 per cent) was not mannite, but reserved judgement on whether the substance was of insect origin (the natives of the region believed not). T. Dobson (1851) reported on the "tent-like protection" of several species of *Psylla*, including *P. eucalypti*.[19] The excretion is the nidus of the larvae or pupae, and is therefore similar in origin to *tréhala* of Persia (*Larinus maculatus* on *Echinops persicus*). F. A. Flückiger and A. Hanbury contributed further chemical analyses,[20] and the substance was exhibited at the International Exhibition in London in 1862.[21]

The host species has usually been identified as *Eucalyptus dumosa* A. Cunn., the "white mallee" or "bull mallee." Other species include *E. mellidora* A. Cunn.,[22] *E. oleosa* F. v. Muell, *E. odorata* Behr. et Schlechtendal, *E. leucoxylum* F. v. Muell,[23] *E. punctata* D. C.,[24] and *E. maculata* Hook.[25] The mallee scrub was sometimes burned at the end of the summer to promote new growth on which the insect thrives. In the earlier literature the *lerp* parasite was described as *Pyslla eucalypti*. R. Basden (1966, 1970) has added *Eurymelia distincta* Signoret (on *E. punctata*)[26] and *Eucalyptolyma maidenii* Froggatt (on *E. maculata*).[27]

(b) Eucalyptus manna

A substance consisting chiefly of melitose (melitriose, raffinose)[28] is found between December and March on the leaves and young branches of members of the genus *Eucalyptus* (Map 17). The exudation is apparently the

---

[16] Beveridge, 1884: p. 64.
[17] Johnston and Cleland, 1942: p. 29.
[18] Anderson (1849), 1851: pp. 241–247.
[19] Dobson, 1851: pp. 235–241. West (1858: p. 75) has a drawing of the insect. See also Wooster, 1882: pp. 91–94.
[20] Flückiger, 1868a: p. 124; 1868b: pp. 161–169; 1871a: pp. 188–190; 1871b: pp. 7–29; 1883: pp. 28–29; Flückiger and Hanbury, 1879: p. 417. Cf. Irving, 1957: p. 140 (sugar 53 per cent, threads [lerp-amylum] 33 per cent).
[21] Hanbury, 1862–1863: pp. 108–109; 1876: p. 283. Cf. Balfour, 1885: 2: p. 853; Maiden, 1889b: p. 508; 1892: p. 2; Wehmer, 1929: p. 536.
[22] Wooster, 1882: p. 92; Chisholm, 1958: p. 480.
[23] Tepper, 1883: p. 109.
[24] Basden, 1966: p. 44.
[25] Basden, 1970: p. 9.
[26] Basden, 1966: p. 44. Basden observed that the sugars of the phloem sap and of *lerp* (70 per cent raffinose) differ and suggested that an enzyme in the saliva of the insect may be responsible.
[27] Basden, 1970: p. 9. Macpherson (1939: p. 180) had earlier recognized that several species of insect were involved in the production of *lerp*.
[28] Thomson, 1838: pp. 640, 642; Johnston, 1843: p. 14; Anderson (1849), 1851: p. 241; Berthelot, 1855: pp. 392–393; Flückiger and Hanbury, 1879: p. 417; Flückiger, 1883: p. 28; McCoy, 1885–1890: 1: p. 55; Passmore, 1891: p. 717; H. G. Smith, 1897: p. 177; Ebert, 1908: pp. 503–504.

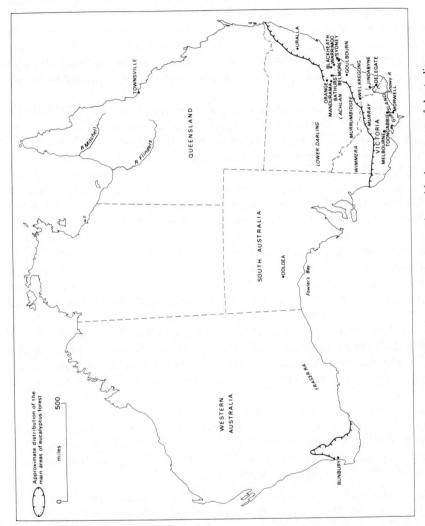

Map 17. Regions and places mentioned in connection with the mannas of Australia.

result of drought (and rupture of the cortical vessels of the tree) and/or of punctures by insects (*Psaltoda* [*Cicada*] *moerens*, among others),[29] and occurs particularly on the sun-facing side of the tree.[30]

One of the earliest reports comes from northern Tasmania. In September 1808, Col. W. Paterson wrote to Sir Joseph Banks: "I have made some discovery of .... an insect which produces very fine manna, which has been given as that medicine and proves equally good. It is only found on the narrow-leaved eucalyptus [? *E. viminalis*], where thousands of these insects resort to about the beginning of November and continue till January in the winged state .... This sachrine (*sic*) substance can be gathered in large quantities; I am certain upwards of 20 pounds might be procured from one tree ...."[31]

The natives of the Goulbourn Plains (New South Wales) informed the naturalist George Bennett (1830's) that "the manna was caused by the *galang-galang*, their name for the tittigonia" (including *C. moerens*).[32] The manna itself (*cú hingaban*) was collected from around the tree (*bartoman*) and eaten.[33] J. L. Stokes (1838) observed two kinds of manna (white and pale yellow) in the vicinity of and in the hill-country to the north of Melbourne. Natives "sometimes scrape from the tree as much as a pound in a quarter of an hour. It has the taste of a delicious sweetmeat ...."[34] G. C. Mundy, when crossing the plains of Bathurst (1846), found "*E. mannifera* or Flooded gum .... in great profusion and of majestic size .... [Manna] is available in small pieces and on the ground under the trees at certain seasons, or in hardened drops on the surface of the leaves; it is snowy white when fresh .... [and] sweeter than the sweetest sugar. The manna is seldom plentiful, for birds and beasts and human beings devour it ...."[35] Most of the available information comes from Tasmania and from south eastern Australia (Victoria and New South Wales), the old Australia Felix (Appendix B). In northern Queensland (Flinders and Mitchell river basins) E. Palmer reported that "Manna is procured from the leaves and small branches [of *E. terminalis*] by being gathered and laid on pieces of bark, when the particles of sugar and gum fall off, or are scraped off with mussel shells into a *kooliman* (bowl), or the leaves when covered with the sweet exudation are pounded together with a stone and roasted in the ashes. After the rainy season this food is said to be abundant."[36] In central Australia acacia gum took the place of eucalyptus manna.[37]

---

[29] McCoy, 1885–1890: 1: p. 55; Ebert, 1908: pp. 427, 503; Maiden, 1920: pp. 108, 111; Myers, 1929: p. 164. The role of insects was first suggested by Johnston, 1843: p. 14.
[30] Discussed by Basden (1965: pp. 153, 155), who also identifies the following insects: *Philactophaga eucalypti, Hyalarcta hubneri,* and *Perga dorsalis.*
[31] Paterson, 1900: p. 768.
[32] G. Bennett, 1860: p. 272.
[33] G. Bennett, 1834: 1: p. 320. Cf. Eyre (1840–1841), 1845: 2: p. 273.
[34] Stokes, 1846: 1: pp. 285–286.
[35] G. C. Mundy, 1855: p. 176.
[36] E. Palmer, 1884: pp. 98–99. Repeated by Roth, 1901: p. 12.
[37] Spencer and Gillen (1938: p. 185) write of the "men of the *ilpirla* [totem]." Ilpirla was "a kind of manna" found on the mulga tree (*Acacia aneura* F. v. Muell). Among the Arunta (around Alice Springs) a ceremony (*intichiuma*) was enacted to increase the supply of *ilpirla*. *Ilpirla* also referred to a drink prepared from the honey ant, *Melophorus* (*Camponotus*) *inflatus*.

Eucalyptus manna, like the insect excretion *lerp*, was probably added to water and allowed to ferment. In Tasmania the saccharine sap of *E. gunnii* Hook. ("cider tree" or "sugar gum") was tapped for this purpose by natives and by early white settlers.[38] The exuded manna has been observed more particularly on *E. viminalis* Labill. (1806), known as *yarra yarra*, "manna gum" and "peppermint gum." J. H. Maiden found however that *E. rubida* Deane and Maiden (*E. mannifera* A. Cunn. and Mudie) yielded more manna than *E. viminalis* in New South Wales.[39] The latter is occasionally "cultivated."[40] Introduced eucalypti in Argentina (*E. viminalis*) and the Nilgiri region of southern India are known to exude a saccharine fluid.[41]

Manna from one or other of these species (and possibly *E. gunnii*) was brought to England from Australia in the first decade of the 19th century, apparently as a substitute for the product of *Fraxinus ornus*.[42] It was displayed at the Paris Exhibition of 1855[43] and again at the London Exhibition of 1862,[44] but exports never reached significant proportions. Eucalyptus manna, like others of Australia and Tasmania, is interesting chiefly as a formerly more important item in the diet of the native population.

---

[38] W. J. Hooker, 1843: no. 4036 (*Eucalyptus* sp.); Macpherson, 1939: p. 178 (River Shannon region); Irving, 1957: p. 140. Bunce (1857: p. 47) identified the cider tree as *E. resinifera* Smith. The "saccharine liquor" was known as *wayalinah*. See also Noetling, 1911: p. 283; and on *E. gunnii*, Passmore, 1891: pp. 717–720.

[39] Maiden, 1909–1933: 3, 8: p. 170. On the synonym *E. mannifera*, see *ibid.*, 3, 6: p. 111. Furthermore, according to Maiden (1920: p. 105), "A good deal of manna referred to *E. viminalis* in the past belongs to *E. rubida* because, until the description of the latter, it was usually looked upon as a form of *E. viminalis*." S. G. Harrison (1951: p. 413) gives *E. viminalis* Labill. (syn. *mannifera* A. Cunn.)

[40] Wehmer, 1929: pp. 534–535.

[41] Ducloux, 1828: pp. 73–76 (from La Plata, Argentina; glucose 15.50 per cent, sacrose 25.30 per cent, raffinose 54 per cent). The eucalyptus was introduced to South America in the early 1820's (Dickinson, 1969: pp. 295–296). For the Nilgiri region, see Dey, 1896: p. 133.

[42] Virey, 1832: p. 706 (*E. mannifera*); Thomson, 1838: pp. 640, 642 (*E. mannifera* from Botany Bay, ca. 1815); Anderson, 1851: p. 241 (*E. mannifera*, ca. 1819); Chisholm, 1958: 5: p. 480 (by 1809). Mudie (1829: p. 155) noted that "the manna obtained from this tree [*E. mannifera*] is described as having the same medicinal properties as that of the manna ash of Italy."

[43] Berthelot, 1855: pp. 312–313 (? *E. gunnii*).

[44] Hanbury, 1862–1863: pp. 108–109 (*E. viminalis*).

7 MANNAS OF THE NEW WORLD

A. MIDDLE AND SOUTH AMERICA

Early references to manna in the New World are few and, for the most part, unspecific. The English traveller Henry Hawks, reporting on the "commodities of *Nova Hispania*" (1572), states that "there is among the wilde people much manna. I have gathered of the same, and have eaten it, and it is good; for the apothecaries send their servants at certaine times, to gather of the same purgations, and other uses."[1] He may refer to the sweet gum of the "honey mesquite," *Prosopis juliflora* D.C., or, more likely, to one or other of the medicinal mannas mentioned by Francisco Hernández in his monumental *Historia natural de Nueva España* (1571–1576). One kind was found on *Ceiba* sp., the Náhuatl *itzámatl*[2] and *pochotli*[3] (*C. pentandra* Gaertn., the "silk cotton tree"), the other, "en nada inferior al que se importa de Campania," on a species of *Salix*.[4]

Antonio de Herrera (ca. 1600) knew of the manna of Mexico (Tlaxcala)[5] and of Chile; the latter he described as falling as a heavy dew (*gran rocío*), collected "como pan de azúcar."[6] *Salix chilensis* (*S. humboldtiana*), according to Ignatius Molina (1787), "yielded annually a great quantity of manna."[7] Other 17th- and 18th-century statements are probably based, wholly or in part, on Herrera.[8]

In Lower California between April and June the Sicilian Jesuit Fr. Francesco María Picolo observed (1702) manna on some kind of reed

---

[1] Hawks, 1904: p. 385.
[2] Hernández, 1959: 1: p. 87 ("También recogí de este árbol, principalmente entre los Hoaxtepecenses [Huaxtepec, Guaxtepec or Oaxtepec, 28 kilometres east of Cuernavaca], el llamado maná por los árabes, muy semejante al nuestro en forma y propiedades, pero un poco más duro y glutinoso.")
[3] Hernández, 1959: 1: p. 320 (*Del Hoeipochotli* – "Cuando abunda el rocío celeste, suele condensarse sobre este género de árboles en el llamado por los árabes maná, pero algo más duro que el nuestro y sumamente glutinoso.") *Pochotl* is described by Bernardino de Sahagún (*Florentine Codex* [ca. 1570]. 1963: p. 108) but without reference to manna.
[4] Hernández, 1959: 1: p. 87 ("En otros lugares de esta Nueva España, como los Quauhnahuacenses [Cuahnáhuac, Cuernavaca], suele cuajar sobre los sauces un maná excelente, en nada inferior al que se importa de Campania.")
[5] Herrera, 1934–1956: 5: p. 98 ("y en algunas partes se coge mana y no maná, suavísimo que sirve para purgar y preparar estómagos debilitados.") Cf. Boyd-Bowman, 1971 (*maná*).
[6] Herrera, 1934–1956: 14: p. 44 ("y en algunos valles, por sus tiempos, cae tan gran rocío, que se cuaxa y coge como pan de azúcar, y es tan saludable que lo llaman maná.")
[7] Molina, 1809: 1: p. 137.
[8] Pomet, 1694: 1: p. 239 ("Les Mexiquains ont de la manne, qu'ils mangent, comme nous faisons le fromage."); Savary des Bruslons (1723), 1742: 2: p. 1184; *Commerce*, 1783–1784: 3: p. 65. Kolb (1892: p. 2) refers to "American manna."

(*roseaux*).[9] This has been tentatively identified as the sweet excretion of aphids on *Phragmites communis* Trin., a substance collected by the Utes into modern times.[10] J.-B. Labat, in a discussion of the mannas of Europe, added that he had found (ca. 1700) a similar product, like "white honey," in the highlands of Guadeloupe and Martinique.[11]

B. NORTH AMERICA

At least three members of the Coniferae of western North America are known to produce manna. One is the sugar pine, *Pinus lambertiana* Dougl., in southern Oregon and northern California,[12] and another the Douglas fir, *Pseudotsuga taxifolia* Britton of British Columbia.[13] The third species is the Western larch, *Larix occidentalis* Nutt., also of British Columbia.[14]

There is an account of the first of these in *Narrative of the United States Exploring Expedition*, 1838–1842: "Some of the sugar produced by [*Pinus lambertiana*] was obtained [1841]; it is of a sweet taste, with a slightly bitter and piny flavour; it resembles manna, and is obtained by the Indians by burning a cavity in the tree, whence it exudes. It is gathered in large quantities. This sugar is a powerful cathartic, and affected all the party that partook of it; yet it is said that it is used as a substitute for sugar among the trappers and hunters."[15]

The manna of *Pseudotsuga taxifolia* is found on the twigs and needles after hot, dry summers and consists of 75 to 83 per cent melezitose. Whether aphids play any part in the production of this substance is apparently still in doubt. So too is its use as food by the Indian population.

---

[9] Picolo, 1715: p. 279 ("Dans les mois d'Avril, de Mai et de Juin, il tombe avec la rosée une espece de Manne, qui se congele et qui s'endurcit sur les feuilles des roseaux, sur lesquelles on la ramasse. J'en ai goûté. Elle est un peu moins blanche que le sucre; mais elle en a toute la douceur.")

[10] Lloyd, 1897: pp. 329–338. Cf. Sturtevant (1919), 1972: p. 430. Hemsley (1879–1888: 3: p. 571) described *P. communis* as "perhaps the most widely diffused of the larger seed-grasses; common in Mexico and Central America." *Cicada mannifera* has been reported from Brasil (Leunis, 1844–1853: 1: p. 319).

[11] Labat, 1730: 5: p. 316. Cf. Labat, 1742: 1: p. 96 ("la gomme Elemi"); 7: pp. 371–393.

[12] Planchon and Collin, 1895–1896: 1: p. 750; Dragendorff, 1898: p. 68; Wehmer, 1929: p. 11.

[13] Hudson and Sherwood, 1918: pp. 1456–1460; Henry, 1924: p. 388. *Pseudotsuga taxifolia* = *P. douglasii* Carr. = *P. menziesii* (Mirb.) Franco.

[14] S. G. Harrison, 1951: p. 410 (found 1898).

[15] Wilkes, 1845: 5: p. 232. The sugar was analysed and named *pinite* by Berthelot, 1856: pp. 157–158.

8 POSTSCRIPT

A partiality for sweet foods is probably an ancient human characteristic. It is found today in all parts of the world and at every cultural level. However such foods are not usually staple items of diet, but rather luxuries, prominent in festivals and celebrations. Manna, like honey, was presumably collected long before the earliest written evidence. Pre-agricultural folk were in a position to take advantage of random accumulations. The practice survived among pastoral nomads and, where opportunity existed, among those who were primarily sedentary farmers.

The nature of the substance was apparently nowhere properly understood until the early medieval period, and until much later it was widely held to be some kind of dew, condensed from the atmosphere. An unpredictable resource, the mystery of its origin invited speculation. The notion of a supernatural "gift" or "bonus" appears to have been adopted throughout and beyond the main areas of supply. To this were added other miraculous properties and associations. The use of manna as food and in medicine may not at first have been distinguishable and still to some extent overlap. These customs and beliefs probably antedate the period of the Exodus and the origin of the most famous of food legends.

Manna is chiefly associated with the seasonally hot and dry lands of western and west-central Asia. Arab and Persian scholars of the Middle Ages supply most of the early information, and interest in the product, or rather group of products, was longest maintained in the highland zone between the Oxus (Amu Darya), the Euphrates and the Indus. The principal manna of Europe, that of *Fraxinus ornus*, may have been first exploited during the period of the Arab occupation of Sicily.

The ecology of production is still often obscure. Insect honeydew and exudations from the branches and leaves of trees and shrubs are not infrequently confused. Moreover the name "manna" has been applied, on account of its wider connotation, to altogether different products, such as several wild grains and the lichen *Lecanora esculenta*. Further research is needed on the mannas of the New World, clearly but unspecifically reported in the 16th- and 17th-century literature.

Honey, also regarded as a kind of dew, was the chief early source of sugar. The preparation of date syrup is first reported from the Tigris-Euphrates lowlands. The ennobled sugar cane spread from India, at a relatively late period, to favourable localities within the manna zone. By comparison with such domesticated products, manna was difficult to collect in quantity and very variable in the amount available between one year and

116

the next. The importance of cane sugar increased, along with advances in refining, from the early Middle Ages. Nevertheless manna continued to be used in medicine and in traditional sweetmeats (the latter mainly in the core region of western and west-central Asia) and simply as a supplementary form of sweetening. Conceivably, collection by children played some part in maintaining interest.

Manna, again like honey, was included in systems of *materia medica* from China to the Atlantic shores of Spain. It was similarly valued in the New World, in the 16th century if not earlier. Where not produced locally, apothecaries' needs were met by trade. Medicinal manna was a substance of relatively high value and thus suited to long-distance movement. From inland entrepôts such as Ispahan, Bukhāra and Kabūl, and through the sea ports of the Levant and the Persian Gulf, this "Oriental drug" was made available in innumerable small towns across the width of Eurasia.

APPENDICES

APPENDIX A

Reports of *Alhagi maurorum* Desv. and *Alhagi camelorum* Fisch. (Map 2).

Alhagi maurorum Desv. (*Hedysarum alhagi* L., A. *mannifera* Desv., A. *karduchorum* Boiss. et Haussk., A. *graecorum* Boiss.)

Afghanistan
Brandis, 1874: p. 144.

Anatolia
Tschihatchcheff, 1853–1869: 3, 1: p. 105; Davis, 1965–1978: 3: p. 597; Guest and Townsend, 1966–1974: 3: p. 502.

Arabia, Palestine, and the Lebanon
Forskål, 1775a: p. 136; Olivier, 1801–1807: 3: p. 189; Délile, 1812: p. 10; Sprengel, 1825–1828: 3: p. 316; Jaubert and Spach, 1842–1857: 5: tab. 401; Boissier, 1867–1888: 2: p. 559; Hart, 1885: p. 430; 1891: p. 91; Volkens, 1887: p. 107; Dragendorff, 1898: p. 326; Dinsmore and Dalman, 1911: p. 605; Blatter, 1919–1933; p. 164; Musil, 1927: p. 589; Post, 1932–1933: 1: p. 415; Davis, 1965–1978: 3: p. 597; Guest and Townsend, 1966–1974: 3: p. 502.

Armenia, Georgia
Tournefort (1717) 1741: 2: p. 4; Don, 1831–1832: 2: p. 310.

Baluchistān
Watt, 1889–1893: 1: p. 165.

Caucasus
Guest and Townsend, 1966–1974: 3: p. 502.

Crete
Olivier, 1801–1807: 3: p. 189.

Cyprus
Olivier, 1801–1807: 3: p. 189; Davis, 1965–1978: 3: p. 597; Guest and Townsend, 1966–1974: 3: p. 500.

Egypt
Forskål, 1775a: p. 136; Délile, 1812: p. 10; Don, 1831–1832: 2: p. 310; Visiani, 1836: p. 33; Boissier, 1867–1888: 2: pp. 558–559; Oliver *et al.*, 1868–1937: 2: p. 142; Brandis, 1874: p. 144; Comes, 1879: p. 5; Volkens, 1887: p. 107; Dragendorff, 1898: p. 326; Schweinfurth, 1912: p. 4; Muschler, 1912: 1: pp. 536–537; Blatter, 1913–1933: p. 164; Davis, 1965–1978: 3: p. 597; Guest and Townsend, 1966–1974: 3: pp. 500, 502.

Greece and the Aegean Islands
Tournefort (1717), 1741: 2: p. 4; Olivier, 1801–1807: 3: p. 189; Sibthorp, 1806–1813: 2: p. 82; 1806–1840: 8: p. 15; Sprengel, 1825–1828: 3: p. 316; Jaubert and Spach, 1842–1857: 5: tab. 401; Brandis, 1874: p. 144; Davis, 1965–1978: 3: p. 597; Guest and Townsend, 1966–1974: 3: p. 500.

Indian Sub-Continent
Roxburgh, 1820–1832: 3: p. 344; Don, 1825: pp. 246–247; Bellew, 1864: p. 238; Aitchison, 1869: p. 44; J. D. Hooker, 1872–1897: 2: p. 145; Brandis, 1874: p. 144; Watt, 1889–1893: 1: p. 165; Cooke, 1903–1908: p. 333; Bamber, 1916: p. 79; Kirtikar and Basu, 1918: 1: pp. 421–422.

Iraq
Rauwolf (1573–1576), 1693: p. 152; Don, 1831–1832: 2: p. 310; Tschihatchcheff, 1853–1869: 3, 1: p. 105; Blatter, 1919–1933: p. 164; Anon., 1929; p. 6; Zohary, 1950: p. 93.

118

Persia
Olivier, 1801–1807: 3: p. 189; Délile, 1812: p. 10; Sprengel, 1825–1828: 3: p. 316; Jaubert and Spach, 1842–1857: 5: tab. 401; Boissier and Buhse, 1860: p. 90; Boissier, 1867–1888: 2: p. 559; Dragendorff, 1898: p. 326; Davis, 1965–1978: 3: p. 597; Guest and Townsend, 1966–1974: 3: p. 502.

Sahara
Duveyrier, 1864: p. 163; Rohlfs, 1875: tab. 6; Muschler, 1912: 1: pp. 536–537; Trotter, 1915: p. 180; Chevalier, 1933: pp. 277–278; Killian, 1947: pp. 52–57; Nicolaisen, 1963: p. 177; Davis, 1965–1978: 3: p. 597; Guest and Townsend, 1966–1974: 3: p. 502; Barth (1850), 1972: glossary (*aghul*); Ozenda (1958), 1977: p. 306.

Syria
Rauwolf (1573–1575), 1755: pp. 93–94; Olivier, 1801–1807: 3: p. 189; Sprengel, 1825–1828: 3: p. 316; Don, 1831–1832: 2: p. 310; Boissier, 1867–1888: 2: p. 559; Oliver *et al.*, 1868–1937: 2: p. 142; Dragendorff, 1898: p. 326; Blatter, 1913–1933: p. 164; Post, 1932–1933: 1: p. 415; Davis, 1965–1978: 3: p. 597; Guest and Townsend, 1966–1974: 3: p. 500.

Alhagi camelorum Fisch. (A. [*Hedysarum*] *pseudo-alhagi* Desv. (M.B.), *A. persarum* Boiss. et Buhse, *A. turcorum* Boiss., *A. kirghisorum* Schrenk.)

Afghanistan
C. Masson, 1843: p. 455; Boissier, 1867–1888: 2: p. 559; Aitchison, 1886–1887: p. 467; 1888–1894: p. 59; Dragendorff, 1898: p. 326; Blatter, 1919–1933: p. 164; Gilliat-Smith and Turril, 1930: 8: p. 376; Zaman, 1951: pp. 64–65; Kitamura, 1960: p. 185.

Anatolia
Tschihatchcheff, 1853–1869: 3, 1: pp. 105–106; Boissier, 1867–1888: 2: p. 559; Kitamura, 1960: p. 185; Davis, 1965–1978: 3: p. 596; Guest and Townsend, 1966–1974: 3: p. 503.

Arabia, Palestine, and the Lebanon
Dinsmore and Dalman, 1911: p. 606; Blatter, 1919–1933: p. 164; Post, 1932–1933: I: p. 415; Guest and Townsend, 1966–1974: 3: p. 503.

Armenia
Takhtadzhiana, 1954–1966: 4: pp. 270–271.

Baluchistān
C. Masson, 1843: p. 455; Aitchison, 1888–1894: p. 59; Burkill, 1909: p. 26; Blatter, 1919–1933: p. 164.

Caucasus
Don, 1831–1832: 2: p. 310; Boissier, 1867–1888: 2: p. 559; Davis, 1965–1978: 3: p. 596; Guest and Townsend, 1966–1974: 3: p. 503.

Central Asia
Don, 1831–1832: 2: p. 310; Burnes, 1834: 2: p. 167; Ledebour, 1842–1853: I: p. 715; Lehmann (1841–1842), 1852: pp. 248–249; Tschihatchcheff, 1853–1869: 3, 1: pp. 105–106; Boissier, 1867–1888: 2: p. 559; Schefer in Nassir, 1881: p. 270; Paulsen, 1912: p. 219; Gilliat-Smith and Turril, 1930: 8: p. 376; Kitamura, 1960: p. 185; Guest and Townsend, 1966–1974: 3: p. 503.

Cyprus
Guest and Townsend, 1966–1974: 3: p. 503.

Egypt
Tschihatchcheff, 1853–1869: 3, 1: pp. 105–106.

Greece
Tschihatchcheff, 1853–1869: 3, 1: pp. 105–106.

Iberia
Don, 1831–1838: 2: p. 310.

Indian Sub-Continent
Blatter, 1913–1933: p. 164; Bamber, 1916: p. 79; Gilliat-Smith and Turril, 1930: 8: p. 376; Guest and Townsend, 1966–1974: 3: p. 503.

Iraq
Guest and Townsend, 1966–1974: 3: p. 503.

North Africa
Guest and Townsend, 1966–1974: 3: p. 502.

Persia
Boissier and Buhse, 1860: p. 76; Boissier, 1867–1888: 2: p. 559; Aitchison, 1886–1887: p.

119

467; Blatter, 1913–1933: p. 164; Gilliat-Smith and Turril, 1930: 8: p. 376; Kitamura, 1960: p. 185; Davis, 1965–1978: 3: p. 596; Guest and Townsend, 1966–1974: 3: p. 503.

Siberia
Lansdell, 1885: 2: p. 632; Kitamura, 1960: p. 185; Davis, 1965–1978: 3: p. 596.

Southern Russia
Ledebour, 1842–1853: 1: p. 715; Keller, 1926: pp. 127, 131–134; Gilliat-Smith and Turril, 1930: 8: p. 376.

Syria
Tschihatchcheff, 1853–1869: 3, 1: pp. 105–106; Boissier, 1867–1888: 2: p. 559; Guest and Townsend, 1966–1974: 3: p. 503.

APPENDIX B

Reports of eucalyptus manna in Australia and Tasmania (Map 17).

Date	Species	Location	Reference
1808	? *E. viminalis* Labill.	Launceston (Tas.)	Paterson, 1900: p. 768
1809			Chisholm, 1958: 5: p. 480
1815	? *E. viminalis* Labill.	east of Mandurama (N.S.W.)	Cambage, 1921: p. 253
1820	? *E. viminalis* Labill.	near Orange (N.S.W.)	Cambage, 1921: p. 253
1829	*E. mannifera* Mudie	Blue Mountains (N.S.W.)	Mudie, 1829: p. 155; 1832–1833: p. 24
1831	*E. mannifera* Mudie	New Holland	Merat and Lens, 1829–1834: 3: p. 173
1832	*E. mannifera* Mudie		Virey, 1832: p. 706
1832-1834	*E. mannifera* Mudie	Goulbourn Plains (N.S.W.)	G. Bennett, 1834: 1: p. 319
1838	*E. mannifera* Mudie	New South Wales	Thomson, 1838: pp. 640,642
1838	*Eucalyptus* spp.	Melbourne region (Vict.)	Stokes, 1846: 1: pp. 285–286
1841	*E. mannifera* Mudie		Eyre, 1845: 2: p. 273
1843	*E. virgata* Sieber ex. D.C. (? *E. viminalis* Labill.)	Tasmania	W. J. Hooker, 1843: no. 4036 (cf. Maiden, 1920: p. 107)
1843	*Eucalyptus* spp.	Tasmania	J. F. W. Johnston, 1843: p. 14
1846	*E. mannifera* Mudie	Bathurst Plains (N.S.W.)	Mundy, 1855: p. 176
1847	*E. mannifera* Mudie	New Holland	Leunis, 1844–1853: 2: p. 284
1849	*E. mannifera* Mudie	New South Wales	Anderson, 1851: p. 241
1853-1854	? *E. resinifera* Smith		Landerer, 1854: p. 412 (cf. Maiden, 1920: p. 104)
1855	? *E. gunnii* Hook.	? Welaregong (N.S.W.)	Berthelot, 1855: pp. 392–393
1857	? *E. resinifera* Smith	Tasmania	Bunce, 1857: p. 47
1860	*E. viminalis* Labill.	Goulbourn Plains (N.S.W.)	G. Bennett, 1860: p. 272
1862	*E. viminalis* Labill.		Hanbury, 1862–1863: pp. 108–109; 1876: p. 283
1870	*Eucalyptus* sp.	Tasmania	Bonwick, 1870: p. 16
1874	*E. viminalis* Labill.		Flückiger and Hanbury, 1879: p. 417
1880	*E. viminalis* Labill.	Victoria, New South Wales, Tasmania, Queensland	McCoy, 1885–1890: 1: p. 55
1883	*E. gomphocephala* D.C.	near Bunbury (W.A.)	Maiden, 1920: p. 104
1883	*E. viminalis* Labill. *E. mannifera* Mudie ? *E. resinifera* Smith		Flückiger, 1883: p. 28

120

1883	*E. terminalis* F. v. Muell	Flinders river, Mitchell river (Queens.)	E. Palmer, 1884: pp. 98–99; Roth, 1901: p. 12
1884	*E. viminalis* Labill.		Mueller, 1884: no pagination
1885	*E. viminalis* Labill. ? *E. resinifera* Smith		Balfour, 1885: 2: p. 853
1889	*E. viminalis* Labill.	Victoria, New South Wales, South Australia, Tasmania	Maiden, 1889a: p. 27
1889	*E. viminalis* Labill.		Maiden, 1889b: p. 510
1890	*E. viminalis* Labill. (? *E. rubida* Deane and Maiden)	Morwell (Vict.)	Howitt, 1890: p. 99 (cf. Maiden, 1920: p. 106)
1890	*E. stuartiana* F. v. Muell	Toongabbie (Vict.)	Howitt, 1890: pp. 99–100
1891	*E. gunnii* Hook.	Jindabyne, Woollandibby (N.S.W.)	Passmore, 1891: p. 717
1891	*E. pulverulenta* Sims (? *E. cinerea* F. v. Muell)	Buckley's Crossing, Snowy river (N.S.W.)	Anon., 1891b: p. 381
1897	*E. punctata* D.C.	Belmore [Sydney] (N.S.W.)	H. G. Smith, 1897: p. 177
1898	*E. mannifera* Mudie *E. viminalis* Labill. *E. gunnii* Hook.	Gippsland (Vict.) Tasmania	Dragendorff, 1898: p. 47
1902	*E. stuartiana* F. v. Muell	Dalgety [Delegate] N.S.W.)	R. G. Smith, 1902: p. 23
1903	*E. punctata* D.C.	Ryde [Sydney]	Anon. 1903: p. 686
1908	*E. viminalis* Labill. *E. mannifera* Mudie ? *E. resinifera* Smith *E. gunnii* Hook. *E. pulverulenta* Sims		Ebert, 1908: pp. 503–504
ca. 1914	*Eucalyptus* sp.	near Fowler's Bay (S.A.)	Bates, 1938: p. 135
1916	*E. viminalis* Labill.	New South Wales	Maiden, 1909–1933: 3, 8: p. 170
	E. rubida Deane and Maiden	New South Wales	
-1920	*E. cinerea* F. v. Muell (*E. nova-anglica* Deane and Maiden)	Uralla (N.S.W.)	Maiden, 1920: p. 103
1920	*E. corymbosa* Smith		Maiden, 1920: pp. 104, 115
	E. eximia Schauer *E. foecunda* Schauer *E. punctata* D.C. *E. terminalis* F. v. Muell	Western Australia Sydney	Maiden, 1920: p. 104 Maiden, 1920: pp. 107, 115
	E. rubida Deane and Maiden *E. rubida* Deane and	Monaro [mountains, south west of Cooma] Dalgety [Delegate] (N.S.W.)	Maiden, 1920: p. 106
1921	Maiden	New South Wales	Cambage, 1921: p. 253
1922	*E. radiata* Sieber ex D.C. (*E. amygdalina* Labill.)	Blackheath (N.S.W.)	Maiden, 1909–1933: 6, 5: p. 250
1929	*E. eximia* Schauer	Warrimoo (N.S.W.)	Maiden, 1909–1933: 8, 3: p. 145
1929	*E. viminalis* Labill.		Myers, 1929: p. 164
1929	*E. mannifera* Mudie *E. viminalis* Labill.		Wehmer, 1929: pp. 534–535
1939	*E. gunnii* Hook.	Shannon river (Tas.)	Macpherson, 1939: pp. 176: 179–180
	E. viminalis Labill.		

1957	*E. viminalis* Labill.	South Australia, New South Wales, Victoria Tasmania	Irving, 1957: p. 140

1965	*E. gunnii* Hook. *E. mannifera* Mudie *E. viminalis* Labill. *E. punctata* D.C. *E. maculata* Hook. *E. citriodora* Hook. *E. tereticornis* Smith	Tasmania	Basden, 1965: p. 152

APPENDIX C

"Manna" or "honeydew" has been briefly reported on the following species:

Acer sp. — Geoffroy, 1741: 2: p. 590; Quer, 1762–1784: 2: pp. 159–160; Ludwig, 1870: p. 44; Boussingault, 1872: p. 218; Vogel, 1976: pp. 132–133

Alnus [*glutinosa* L.] — Boussingault, 1872: p. 218

Carduus arcticoides Willd. — Darwin, 1876: p. 402 (quoting Treviranus)

Celastrus sp. — Ehrenberg, 1927b: p. 75

Citrus ? aurantium L. — De La Hire, 1709: p. 69

Citrus limon Burmann — De La Hire, 1709: p. 69

Ensete superbum (Roxb.) Cheesman (*Musa superba* Roxb.) — Watt, 1889–1893: 5: p. 165

Euonymus europaeus L. — Wehmer, 1929: p. 455

Fagus ? sylvatica L. — Ludwig, 1870: p. 44 (quoting Treviranus)

Ficus ? carica L. — Geoffroy, 1741: 2: p. 590

Juglans sp. — Haller, 1768: 2: pp. 294–295; Woodville, 1790–1794: 1: p. 105; Langlois, 1843b: pp. 348–351; Ludwig, 1870: p. 44 (quoting Treviranus)

Juniperus sp. — Geoffroy, 1741: 2: p. 590; Sestini, 1788: p. 92

Picea sp. and/or *Abies* sp. — Geoffroy, 1741: 2: p. 590; Engeström, 1776: pp. 144–150; Sestini, 1788: p. 92; Woodville, 1790–1794: 1: p. 105; Musset, 1879: pp. 306–307 (cf. Henry, 1924: p. 388)

Pyrus glabra Boiss. — Haussknecht, 1870: p. 248; followed by Boissier, 1867–1888: 4: pp. 411–412; Flückiger and Hanbury, 1879: p. 416; Dragendorff, 1898: pp. 275, 603; Wood and Bache, 1907: p. 763

Populus sp. — Sestini, 1788: p. 92; Ludwig, 1870: p. 44 (quoting Treviranus); Henry, 1924: pp. 387, 390

Prunus sp. — Boussingault, 1872: p. 218

Rhododendron arboreum Sm. — Watt, 1889–1893: 3: p. 443; 5: p. 165

Rosa sp. — Boussingault, 1872: p. 218

Schrophularia frigida Boiss. — Haussknecht, 1870: p. 248; followed by Boissier, 1867–1888: 4: pp. 411–412; Flückiger and Hanbury, 1879: p. 416; Dragendorff, 1898: pp. 275, 603; Wood and Bache, 1907: p. 763

Ulmus sp. — Ludwig, 1870: p. 44 (quoting Treviranus)

BIBLIOGRAPHY OF WORKS CITED

Abbayes, Henry des. 1951. *Traité de Lichenologie* (New York).

Abbe, E. 1965. *The plants of Virgil's Georgics* (Ithaca).

'Abd ar-Razzāq. 1874. *Kachef ar-Roumoûz* [*Révélation des Énigmes, ou Traité de Matière Médicale Arabe*] (trans. L. Leclerc, Paris).

Abū al-Fidā. 1848–1883. *Géographie d'Aboulféda* (trans. J. T. Reinaud, 2 v., Paris).

Acharius, Erik. 1803. *Methodus qua omnes detectos Lichenes* (2 v., Stockholmiae).

—— 1806. "Arthonia, Novum Genus Lichenum." *Schrader's Neues Journal für die Botanik* 1, 3: p. 22.

—— 1810. *Lichenographia Universalis* (Gottingae).

—— 1814. *Synopsis Methodica Lichenum* (Lundae).

Achart, S. Dj. 1905. *Quinze cents plantes dans l'Inde* (Pondichéry).

Acloque, A. 1893. *Les Lichens* (Paris).

Acosta, Christovão. 1578. *Tractado de las Drogas y Medicinas de las Indias Orientalis* (Burgos).

—— 1585. *Della Historia, Natura, et Virtu delle Droghe Medicinali, et altri Semplici rarissimi, che vengono partati dalle Indie Orientali in Europa* (Venetia).

Actuarius, Ioan [Johannes]. 1556. *Methodi Medendi* (Paris).

Adanson, Michel. 1763. *Familles des Plantes* (2 parties, Paris).

Aelianus, Claudius. 1958–1969. *Aelian: On the characteristics of animals* (ed. and trans. A. F. Scholfield, 3 v., Loeb Classics, London).

Ainslie, W. 1826. *Materia Indica* (2 v., London).

Ainsworth, W. 1868. *Report of a journey from Baghdad to Constantinople, via Kurdistan* [1837]. In: F. R. Chesney *Narrative of the Euphrates Expedition* [1835–1837] (London), pp. 492–541.

Aitchison, J. E. T. 1869. *A Catalogue of the Plants of the Punjab and Sindh* (London).

—— 1886–1887. "Some plants of Afghanistan and their medicinal products." *The Pharmaceutical Journal and Transactions* 3rd. ser., 17: pp. 465–468.

—— 1888–1894. "The botany of the Afghan Delimitation Commission." *Transactions of the Linnean Society* 2nd. ser., 3, 1: pp. 1–139.

—— 1891. "Notes to assist in a farther knowledge of the products of western Afghanistan and of north-eastern Persia." *Transactions of the Botanical Society of Edinburgh* 18: pp. 1–228.

Aiton, W. T. 1810–1813. *Hortus Kewensis* (5 v., 2nd., London).

Al-Bakrī, Abū 'Ubaid. 1913. *Description de l'Afrique septentrionale* (trans. W. Macguckin de Slane, 2nd ed., Alger).

Alberti, Leandro, 1551. *Descrittione di tutta Italia* (Vinegia).

Albertus Magnus. 1867. *De Vegetabilibus* (ed. C. Jessen, Berolini).

Al-Bīrūnī, Abū Raihān. 1973. *Kitāb al-Ṣaydanah: Al-Bīrūnī's book on pharmacy and materia medica* (ed. and trans. H. M. Said and R. E. Elahie, commentary S. K. Hamarneh, 2 v., Karachi).

Al-Dīnawarī, Abū Ḥanīfā. 1953. *The Book of Plants* (ed. B. Lewin, Uppsala Universitets Årsskrift 10, Uppsala).

Alëkhine, (M.) 1889. "Recherches sur le mélézitose." *Annales de Chimie et de Physique* 6ième sér., 18: pp. 532–551.

[Alexandre, Nicolas] 1716. *Dictionnaire botanique et pharmaceutique* (Paris).

Al-Ghāfiqi, Aḥmad ibn Muḥammad. 1932–1938. *The abridged version of "The Book of Simple Drugs" of Aḥmad ibn Muḥammad al-Ghāfiqi by Gregorius Abu'l-Farag* [1226–1286]. (ed. and trans. M. Meyerhof and G. P. Sobhy, 3 v., Cairo).

Alibert, J. L. 1814. *Nouveaux éléments de thérapeutique* (2 v., 3ième ed., Paris).

Al-Idrīsī, Al-Sharīf Muḥammad. 1836–1840. *Géographie d'Édrisi* (trans. P. Amédée Jaubert, 2 v., Paris).

Al-Iṣṭakhrī, Ibrāhīm al-Fārisī. 1845. *Das Buch der Länder* (trans. A. D. Mordtmann, Hamburg).

Allemagne, H. R. d'. 1911. *Du Khorassan au pays des Backhtiaris* (4 v., Paris).

Al-Muqaddasī, Muḥammad ibn Aḥmad. 1892. *Description of Syria, including Palestine* (trans. G. le Strange, London).

—— 1901. *Aḥsanu-T-Taqāsīm Fī MaʿRifatl-L-Aqālīm* (ed. and trans. G. S. A. Ranking and R. F. Azoo, Bibliotheca Indica, Asiatic Society of Bengal, new series, no. 1001, Calcutta).

Alpinus, Prosper. 1591. *De Medicina Aegyptiorum* (Venice).

—— 1592. *De Plantis Aegypti* (Venetiis).

—— 1640. *De Plantis Aegypti Liber cum Observationibus et Notis Ioannis Veslingii* (Patavii).

Al-Rāzī [Rhazes], Abū Bakr Muḥammad. 1766. *Rhazes de Variolis et Morbillis: Arabice et Latine* (ed. and trans. J. Channing, London).

Al-Samarqandī, Muḥammad. 1967. *The Medical Formulary [Aqrābādhīn] of Al-Samarqandī* (ed. and trans. M. Levey and N. Al-Khaledy, Philadelphia).

Al-Ṭabarī, ʿAli ibn Sahl Rabbān. 1969. *Die pflanzliche und mineralische Materia Medica im Firdaus al-Ḥikma des Ṭabarī* (ed. and trans. W. Schmucker, Bonner Orientalistische Studien, Neue Serie 18, Bonn).

Altomarus, Donatus Antonius. 1561. *Medici atque Philosophi Neapolitani* (Venetiis).

—— 1562. *De mannae differentiis, ac viribus, deque eas dignoscendi via, ac ratione* (Venetiis).

Amico e Statella, Vito Maria. 1757–1760. *Lexicon Topographicum Siculum* (3 v., Panormi, Catanae).

Ammann, Johann. 1739. *Stirpium Rariorum in Imperio Rutheno: Icones et Descriptiones* (Petropoli).

Amoreux, P. J. 1787. *Recherches et expériences sur les divers lichens* (Académie des Sciences, Belles-Lettres et Arts. Mémoires sur l'utilité des lichens. Partie 2, Lyon).

Anderson, T. 1851. "On a new species of manna from New South Wales." *Papers and Proceedings of the Royal Society of Van Dieman's Land* 1: pp. 241–247.

André, J. 1956. *Lexique des termes de botanique en Latin.* Études et Commentaires, 23 (Paris).

Andreu, R. F. 1953–1955. "El Mana en la Historia." *Farmacognosía* 13, 28: pp. 83–89, 13, 29: pp. 187–222; 14, 32: pp. 115–146; 15, 38: pp. 303–331.

Angelus, a Sancto Joseph. 1681. *Pharmacopoea persica ex idiomate persico in latinum conversa* (Lutetiae).

Anon. 1828. "Manna erzeugende Insekten." *Archiv des deutschen Apotheker-Vereins in nördlich Deutschland [Archiv der Pharmazie]* 24: p. 262.

—— 1847. [Report to the government of Algiers] *Athenaeum* no. 1031: p. 816.

—— 1891a. [Note on a fall of *Lecanora esculenta*] *Nature* January 15th., 1891: p. 255.

—— 1891b. [Note on manna] *Proceedings of the Linnean Society of New South Wales* 2nd. ser., 6: p. 381.

—— 1903. [Note on manna] *Proceedings of the Linnean Society of New South Wales* 2nd. ser., 28: p. 686.

—— 1913. "A new variety of "manna" from Rhodesia." *Bulletin of the Imperial Institute* 11: pp. 332–333.

——1927. [Report on the Bodenheimer expedition to Sinai] *Chemist and Druggist* 107: pp. 429, 471.

—— 1929. *Names of Plants of Iraq* (Department of Agriculture, Baghdad).

—— 1937. *Ḥudūd al-ʿĀlam: The Regions of the World* (ed. and trans. V. Minorsky, Oxford).

—— 1970. "La Manne: sa légende, sa nature. Son utilisation médicinale." *Journal des Medecins du Nord et de l'Est*, 7, 6: pp. 108–113.

Antonio da Uzzano, Giovanni di. 1766. *Libro di Gabelle* [ca. 1442] In: Anon. [Giovanni Francesco Pagnini della Ventura] *Della decima . . . . di Firenze* (4 v., 1765–1766, Lisbona e Lucca) 4: pp. 1–195.

Antoninus, Placentinus [Martyr]. 1849. *Itinerarium* In: J. P. Migne *Patrologiae Cursus Completus* (Paris) 72: pp. 898–948.

Apping, G. 1885. *Untersuchungen über die trehalamanna* (Dorpat).

Aristotle. 1961–1970. *Historia Animalium* (ed. and trans. A. L. Peck, Loeb Classics, 3 v., London).

—— 1963. *Minor Works* (ed. and trans. W. S. Hett, Loeb Classics, London).

Arnold, A. 1877. *Through Persia by Caravan* (2 v., London).

Arnold, F. 1897. "Flechten auf dem Ararat." *Bulletin de l'Herbier Boissier* 5: pp. 631–633.

Ascherson, P. F. A. 1864. *Flora der Provinz Brandenburg* (Berlin).

—— 1895–1896. "Eine verschollene Getreideart." *Brandenburgia: Monatsblatt der Gesellschaft für Heimatkunde der Provinz Brandenburg zu Berlin* 4: pp. 37–60.

Ascherson, P. F. A. and P. Graebner. 1898–1902. *Synopsis der Mitteleuropäischen Flora* (2 v., Leipzig).

Athenaeus. 1957–1967. *The Deipnosphists* (ed. and trans. C. B. Gulick, Loeb Classics, 7 v.,

London).

Aucher-Éloy, P. M. R. 1843. *Relations des voyages en Orient de* 1830 à 1838 (revues et annotées M. le Comte Jaubert, 2 parties, Paris).

Averroës, Abū-l-Walīd [Ibn Rushd]. 1531. *De Simplicibus* (Argentorati).

Bābur [Zahir Ud-Din Muḥammad]. 1826. *Memoirs of Zehir-Ed-Din Muhammed Baber, Emperor of Hindustan* (ed. and trans. J. Leyden and W. Erskine, London).

—— 1912–1921. *The Bâbur-nâma in English* (2 v., ed. and trans. A. S. Beveridge, London).

Bach, M. 1857. "Die Insektenwelt: A – Nahrungsmittel." *Natur und Offenbarung* 34: pp. 289–299.

Backhouse, J. 1843. *Narrative of a visit to the Australian colonies* (London).

Bacon, Francis. 1627. *Sylva Sylvarum or A Naturall Historie* (published by William Rowley, London).

Baillon, M. H. 1871–1892. *Dictionnaire de Botanique* (4 v., Paris).

Baker, R. T., and H. G. Smith. 1897. "On the presence of true manna on a "blue grass," *Andropogon annulatus* Forsk." *Journal and Proceedings of the Royal Society of New South Wales* 30: pp. 291–308.

Balfour, E. 1885. *The Cyclopaedia of India and of eastern and southern Asia* (3 v., 3rd ed., London).

Ball, J. 1878. "Spicilegium Florae Maroccanae." *Journal of the Linnean Society (Botany)* 16: 281–772.

Bamber, C. J. 1916. *Plants of the Punjab. A descriptive key to the flora of the Punjab, north-west frontier province and Kashmir* (Lahore).

Barbey, C. and W. 1882. *Herborisations au Levant, Egypte, Syrie et Mediterranée* (Lausanne).

Barlow, Roger. 1932. *A Brief Summe of Geographie by Roger Barlow* (ed. E. G. R. Taylor, Hakluyt Society, 2nd ser., LXIX, London).

Barth, H. 1972. *Chez les Touaregs de l'Aïr* [1850] (Niamey).

Bartholomaeus Anglicus. 1535. *De Proprietatibus Rerum* (trans. John Trevisa [1398–1399], Londini).

—— 1975–1976. *On the Properties of Things: John Trevisa's translation of Bartholomaeus Anglicus de Proprietatibus Rerum* (2 v., Oxford).

Basden, R. 1965. "The occurrence and composition of manna in eucalyptus and angophora." *Proceedings of the Linnean Society of New South Wales* 90: pp. 152–156.

—— 1966. "The composition, occurrence and origin of lerp, the surgary secretion of *Eurymela distincta* Signoret." *Proceedings of the Linnean Society of New South Wales* 91: pp. 44–46.

—— 1968. "The occurrence and composition of sugars in the honeydew of *Eriococcus coriaceus* Mask." *Proceedings of the Linnean Society of New South Wales* 92: pp. 222–226.

—— 1970. "A note on the composition of the lerp and honeydew of *Eucalyptolyma maidenii* Froggatt." *Proceedings of the Linnean Society of New South Wales* 95: pp. 9–10.

Basiner, T. F. J. 1848. *Reise durch die Kirgisensteppe nach Chiwa*. Beiträge zur Kenntniss des Russischen Reiches 15 (St. Petersburg).

Basset, R. 1899. "Les noms berbères des plantes dans le traité des simples d'Ibn el Beiṭār." *Giornale Società Asiatica Italiana* 12: pp. 53–66.

Bates, Daisy. 1938. *The passing of the aborigines* (London).

Battandier, J. A. 1901. "Production abondante de manne par des oliviers." *Journal de Pharmacie et de Chimie* 6ième sér. 13: pp. 177–179.

Bauhin, Gaspard. 1971. *Pinax theatri botanici* [earlier ed. 1623] (Basileae).

Bauhin, Johann. 1650–1651. *Historiae plantarum universalis* (3 v., Ebroduni).

Baxter, J. H. and C. Johnson (comp.). 1934. *Medieval Latin Word List* (London).

Bedevian, A. K. 1936. *Illustrated polyglottic dictionary of plant names in Latin, Arabic, Armenian, English, French, German, Italian and Turkish languages* (Cairo).

Beek, G. W. Van. 1958a. "Frankincense and myrrh in ancient South Arabia." *Journal of the American Oriental Society* 78, 3: pp. 141–144.

—— 1958b. "Ancient frankincense-producing areas." In: R. L. Bowen and F. P. Albright *Archaeological discoveries in South Arabia* (Baltimore), pp. 139–142.

Bell, R. (trans.) 1937. *The Qur'ān* (2 v., Edinburgh).

Bellew, H. W. 1862. *Journal of a political mission to Afghanistan in* 1857 (London).

—— 1864. *A general report on the Yusufzais* (Lahore).

—— 1875. *The history of Káshgharia* (Calcutta).

Belon du Mons, Pierre. 1553. *De arboribus coniferis, resiniferis ....* (Parisiis).

—— 1555. *Les observations de plusieurs singularitez et choses mémorables trouvées en Grèce, Asie, Judée, Egypte, Arabie et autres pays estranges, rédigées en trois livres* (Paris).

Bennett, G. 1834. *Wanderings in New South Wales* (2 v., London).

—— 1860. *Gatherings of a naturalist in Australia* (London).

Bennett, K. H. 1883. "On *Myoporum platycarpum*, a resin-producing tree of the interior of New South Wales." *Proceedings of the Linnean Society of New South Wales* 7: pp. 349–351.

Bentley, R. and H. Trimen. 1880. *Medicinal Plants* (4 v., London).

Berchtold, Friedrich Von. 1836–1842. *Ökonomisch-technische Flora Böhmens* (3 v., Prag).

Berendes, J. 1965. *Die Pharmacie bei den alten Culturvölkern* (2 v., 2nd ed., Hildesheim).

Berhaut, J. 1967. *Flore du Sénégal* (2nd ed., Dakar).

[Berkeley, M. J.] 1849–1864. "Manna." *Gardeners' Chronicle* 1849: pp. 581, 611–612; 1856: p. 84; 1864: pp. 769–770, 794.

Berkeley, M. J. 1857. *Introduction to Cryptogamic Botany* (London).

Berthelot, M. 1855. "Sur quelques matières sucrées." *Comptes Rendus Hebdomaires des Séances de l'Académie des Sciences* (Paris) 41: pp. 392–396.

—— 1856. "On some saccharine substances." *American Journal of Pharmacy* 28: pp. 156–158.

—— 1859a. "Nouvelles recherches sur les corps analogues au sucre de canne." *Annales de Chimie et de Physique* 3ième sér. 55: pp. 269–296.

—— 1859b. "On melezitose, a new species of sugar." *American Journal of Pharmacy* 31: pp. 61–64.

—— 1863. "Sur la manne du Sinai et sur la manne de Syrie." *Annales de Chimie et de Physique* 3ième sér. 67: pp. 82–86.

Bertin, J., J.-J. Hémardinquer, M. Keul, and W. G. L. Randles. 1971. *Atlas des Cultures Vivières* (Paris, The Hague).

Beveridge, P. 1884. "On the aborigines inhabiting the great lacustrine and riverine depression of the Lower Murray, Lower Murrumbidgee, Lower Lachlan, and Lower Darling." *Proceedings of the Royal Society of New South Wales* 17: pp. 19–85.

Beyerlin, W. 1966. *Origins and history of the oldest Sinaitic traditions* (trans. S. Rudman, Oxford).

Bhishagratna, Kaviraj Kunja Lal (ed. and trans.). 1907–1918. *The Sushruta Samhita* (4 v., Calcutta).

Bieberstein, L. B. F. Marschall von. 1808–1819. *Flora Taurico-Caucasica* (3 v., Charkouiae).

Binning, R. B. M. 1857. *A journal of two years travel in Persia . . .* (2 v., London).

Birdwood, G. 1871. "On the Genus *Boswellia*, with description and figures of three new species." *Transactions of the Linnean Society* 27: pp. 111–148.

Birdwood, G. and W. Foster (eds.). 1893. *The First Letter Book of the East India Company, 1600–1619* (London).

Bishop, I. L. 1891. *Journeys in Persia and Kurdistan* (2 v., London).

Blanchard, R. 1883. *Les coccidés utiles* (Paris).

Blatter, E. 1914–1916. *Flora of Aden*. Records of the Botanical Survey of India 7, 1–3 (Calcutta).

—— 1919–1933. *Flora Arabica*. Records of the Botanical Survey of India 8, 1–5 (Calcutta).

Blatter, E. and F. Hallberg. 1919. "The Flora of the Indian Desert: II." *Journal of the Bombay Natural History Society* 26: pp. 525–551.

Blunt, A. 1879. *Bedouin Tribes of the Euphrates* (2 v., London).

Boccone, Paolo [Silvio]. 1697a. *Museo di Fisica e di Esperienze* (Venetia).

—— 1697b. *Curiöse Anmerckungen über ein und ander natürliche Dinge* (Frankfurt und Leipzig).

Bochart, Samuel. 1663. *Hierozoicon sive Bipertitum Opus de Animalibus Sacrae Scripturae* (2 v., London).

—— 1692. *Opera Omnia* (3 v., Lugduni Batavorum).

Bodaeus à Stapel, Joannes. 1644. *Theophrastes: Historia Plantarum* (Amstelodami).

Bodenheimer, F. S. 1928–1929. *Materialen zur Geschichte der Entomologie bis Linné* (2 v., Berlin).

—— 1937. *Prodromus Faunae Palestinae* (Cairo).

—— 1947. "The Manna of Sinai." *Biblical Archaeologist* 10: pp. 2–6.

Bodenheimer, F. S. and O. Theodor. 1929. *Ergebnisse der Sinai-expedition der Hebräischen Universität, Jerusalem, 1927* (Leipzig).

Boissier, E. P. 1839–1845. *Voyage botanique dans le midi de l'Espagne pendant l'année 1837* (2 v., Paris).

—— 1867–1888. *Flora Orientalis* (5 v., and supplement, Geneva and Basle).

Boissier, E. P. and F. A. Buhse. 1860. *Aufzaehlung der auf einer Reise durch Transkaukasien und Persien gesammelten Pflanzen* (Moskau).

Bonar, H. 1857. *Desert of Sinai* (London).

Bonastre, (M.). 1833. "Comparatif entre la manne dite de Briançon et celle du fraxinus excelsior." *Journal de Pharmacie* 19: pp. 443–447.

Bonnier, G. 1896. "Recherches expérimentales sur la miellée." *Revue Générale de Botanique* 8: pp. 1–22.

Bonwick, J. 1870. *Daily life and origin of the Tasmanians* (London).

Bor, N. L. 1953. *Manual of Indian Forest Botany* (Oxford).

Borgen, P. 1965. *Bread from Heaven: an Exegetical Study of the concept of manna in the Gospel of John and the writings of Philo.* Supplement to Vetus Testamentum 10 (Leiden).

Bornmüller, J. F. N. 1906–1907. "Plantae Straussianae sive enumeratio plantarum a Th. Strauss annis 1889–1899 in Persia occidentale collectarum." *Beihefte zum Botanisches Centralblatt* 19: pp. 195–270; 20: pp. 151–196; 22: pp. 102–142.

Boucheman, A. De. 1934. *Matériel de la vie bédouine, recueilli dans le désert de Syrie* (Damas).

Bougon (Dr). 1898. "La manne des Hébreux dans le désert." *Le Naturaliste* 2ième sér. 20: pp. 41–42.

Bourlier, Ch. 1857. "Thrane ou Tricala." *Revue Pharmaceutique* 1857: pp. 37–38.

Boussingault (M.). 1872. "Sur une matière sucré apparue sur les feuilles d'un tilleul." *Journal de Pharmacie et de Chimie* 4ième ser. 15: pp. 214–218.

Bové, N. 1834. "Relation abrégée d'un voyage botanique en Égypte, dans les trois Arabies, en Palestine et en Syrie." *Annales des Sciences Naturelles* 2ième sér. 1: pp. 161–179.

—— 1835. [Report on manna] *Nouvelles Annales des Voyages et des Sciences Géographiques* 68, 4: p. 260.

Brandis, D. 1874. *The forest flora of north-west and central India* (London).

Brant, J. 1841. "Notes of a journey through part of Kurdistan in the summer of 1838." *Journal of the Royal Geographical Society* 10: pp. 341–432.

Brasavolus, Antonius Musa. 1537. *Examen omnium simplicium medicamentorum* (Lugduni).

Bretschneider, E. 1871. *On the knowledge possessed by the ancient Chinese of the Arabs and Arabian colonies and other western countries* (London).

—— 1882–1895. *Botanicon Sinicum* (3 v., London, Shanghai).

—— 1888. *Mediaeval researches from eastern Asiatic sources* (2 v., London).

—— 1898. *History of European botanical discoveries in China* (London).

Breydenbach, Bernhard Von. 1486. *Peregrinatio in Terram Sanctam* (In civitate Morguntina).

—— 1502. *Sanctarum Peregrinationum* (Speier).

—— 1905. *Deane Breidenbach's journey. 1483.* In: Samuel Purchas *Hakluytus Posthumus or Purchas His Pilgrimes* (20 v., 1905–1907, Glasgow) 8: pp. 359–373.

Brives, A. 1909. *Voyages au Maroc, 1901–1907* (2 v., Alger).

Brockbank, O. [? 1920]. *Diary of a journey through the Sinai peninsula and Arabia in 1914* (no place of publication).

Brockelmann, C. 1923–1928. *Lexicon Syriacum* (2nd ed., Gottingae).

Brothwell, D. and P. 1969. *Food in Antiquity* (London).

Brown, R. 1810. *Prodromus Florae Novae Holandiae et Insulae Van-Diemen* (London).

—— 1830. *Prodromus Florae Novae Holandiae et Insulae Van-Diemen: Supplement* (London).

Bruce, James. 1970. *Travels to discover the source of the Nile, in the years 1768 to 1773* (5 v., Edinburgh).

Brugsch, H. 1867–1880. *Hieroglyphisch-Demotisches Wörterbuch* (7 v. in 3, Leipzig).

Brugsch, H. and J. Dümichen. 1862–1865. *Recueil de monuments égyptiens* (7 v., Leipzig).

Bruz, Ladislaus. 1775. *Dissertatio inauguralis de gramine mannae, sive festuca fluitante* (Viennae).

Brydone, P. 1773. *Tour through Sicily and Malta* (2 v., London).

Budge, E. A. T. W. 1913. *The Syriac Book of Medicines* (Oxford).

Bunce, D. 1857. *Twenty-three years wanderings in Australia and Tasmania* (Geelong).

Bunge, Alexander Von. 1835. *Enumeratio plantarum, quas in China boreali collegit. Anno 1831.* Mémoires présentés à l'Académie Impériale des Sciences de Saint-Pétersbourg par divers savans, 2 (St. Pétersbourg).

—— 1852. *Tentamen generis tamaricum species accuratius definiendi* (Dorpat).

Burchard. 1896. *A description of the Holy Land by Burchard of Mount Sion, A.D. 1280* (ed. and trans. A. Stewart, London).

Burckhardt, Johann Ludwig. 1822. *Travels in Syria and the Holy Land* (London).

—— 1829. *Travels in Arabia* (2 v., London).

—— 1830–1831. *Notes on the Bedouins and Wahabys, collected during travels in the East* (2 v., London).

Burkill, I. H. 1909. *A working list of the flowering plants of Baluchistan* (Calcutta).

Burkill, I. H. and Mohamed Haniff. 1930. "Malay village medicine." *The Gardens' Bulletin* (Straits Settlements) 6, 10–11: pp. 165–321.

127

Burnaby, F. G. 1877. *On horseback through Asia Minor* (2 v., London).

Burnes, A. 1834. *Travels into Bokhara* (3 v., London).

Büsching, A. F. 1775. "Gesammelte Nachrichten von dem morganländischen Manna." *Wöchentlichen Nachrichten von neuen Landcharten* (Berlin) 3: pp. 41–48.

Buxtorf, Johann. 1747. *Dissertatio de Manna*. In: Blasio Ugolino (ed.) *Thesaurus Antiquitatum Sacrarum* (34 v., 1744–1769, Venetiis) 8: cols. 587–640.

Cadet (M.). 1810. "Manne observée sur un saule." *Bulletin de Pharmacie* 2: p. 130.

Callcott, M. 1842. *Scripture Herbal* (London).

Cambage, R. H. 1921. "Exploration between Wingecarribee, Shoalhaven, Macquarie and Murrumbidgee rivers (New South Wales)." *Journal and Proceedings of the Royal Australian Historical Society* 7: pp. 217–288.

Camus, A. 1934–1948. *Les Chênes: monographie du genre Quercus* (5 v. [1–2 texte, 3–5 atlas] Paris).

Candolle, A. de. 1835. *Introduction à l'étude de la botanique* (2 v., Paris).

Carnoy, A. J. 1959. *Dictionnaire étymologique des noms grecs de plantes* (Louvain).

Cartwright, John. 1611. *The Preachers Travels* (London).

Castanhoso, Miguel de. 1902. *The Portuguese expedition to Abyssinia in 1541–1542, as narrated by Castanhoso* (ed. and trans. R. S. Whiteway, Hakluyt Society, second series, X, London).

Castela, Henri. 1603. *Le Saint Voyage de Jérusalem et du mont Sinai, fait en 1600* (Paris).

Chang, K. C. (ed.). 1977. *Food in Chinese culture: anthropological and historical perspectives* (New Haven and London).

Chardin, John. 1811. *Voyages du Chevalier Chardin en Perse et autres lieux de l'Orient* (ed. L. Langlès, 10 v., Paris).

—— 1927. *Travels in Persia* (ed. P. Sykes, London).

Chassinat, É. (ed. and trans.). 1921. *Un Papyrus Médical Copte*. Mémoires publiés par les Membres de l'Institut Français d'Archeologie Orientale du Caire, 32 (Le Caire).

Chavannes, E. and P. Pelliot. 1911–1913. "Un traité manichéen retrouvé en Chine." *Journal Asiatique* 10ième sér. 18: pp. 499–617; 11ième sér. 1: pp. 99–199, 261–394.

Chevalier, A. 1933. "Les Alhagi producteurs de manne et spécialement ceux du Sahara." *Revue de botanique appliquée et l'agriculture coloniale* 13: pp. 275–281.

Chisholm, A. H. (ed.). 1958. *The Australian Encyclopaedia* (10 v., Sydney, London).

Chopra, R. N. 1933. *Indigenous drugs of India* (Calcutta).

Choron (M.). 1847. "Communiqué les extraits suivant de lettres qui lui [M. Thenard] ont été adressées par M. Choron, professeur au collége de Saint-Denis (île Bourbon)." *Comptes Rendus Hebdomadaires des Séances de l'Académie des Sciences* (Paris) 25: p. 397.

Chun, W. Y. n. d. *Chinese Economic Trees* (Shanghai).

Cirillo, Dominico. 1771. [An account of the manna tree, 1766] *Philosophical Transactions of the Royal Society of London* 60: pp. 233–236.

Clarke, E. D. 1810–1823. *Travels in various countries of Europe, Asia, and Africa* (6 v., Cambridge and London).

Cleghorn, H. 1870. "Notes on the botany and agriculture of Malta and Sicily." *Transactions of the Botanical Society* (Edinburgh) 10: pp. 106–139.

Cleland, J. B. and T. H. Johnston. 1937–1938. "Notes on native names and uses of plants in the Musgrave Ranges region." *Oceania* 8: pp. 208–215, 328–342.

Clément-Mullet, J. J. 1870. "Études sur les noms arabes de diverses familles de végétaux." *Journal Asiatique* 15: pp. 5–150.

Clusius, Carolus. 1576. *Rariorum aliquot stirpium per Hispanias obseruatarum historiae, libris dvobvs expressa* (Antverpiae).

—— 1605. *Exoticorum libri decem* (Leyden).

Colebrook, H. T. 1807. "On olibanum or frankincense." *Asiatic Researches* 9: pp. 377–382.

Collin, M. E. 1890. "La matière médicale de la Perse." *Journal de Pharmacie et de Chimie* 1890: pp. 102–108.

Comes, O. 1879. *Catalogo delle Piante raccolte dal Prof. A. Costa in Egitto e Palestina nel 1874* (Napoli).

Commerce. 1783–1784. In: *Encyclopédie méthodique* (190 v., 1782–1830) 3 v., Liége and Paris.

Connor, E. 1698. *The History of Poland* (2 v., London).

Cooke, T. 1903–1908. *The Flora of the Presidency of Bombay* (2 v., London).

Coppens, J. 1960. "Les traditions relatives à la manne dans Exode XVI." *Estudios Eclesiasticos* 34: pp. 473–489.

Cordus, Valerius. 1599. *Dispensatorium* [1535] (Lugduno-Batavorum).

Cosmas Indicopleustes. 1897. *The Christian Topography* (ed. and trans. J. W. McCrindle, Hakluyt Society, XCVIII, London).

128

Cosson, M. E. 1857. "Liste des plantes observées par M. le Dr Reboud dans le Sahara Algérien, pendant l'expédition de 1857 de Laghouat à Ouargla." *Bulletin de la Société Botanique* 4: pp. 469–473.

Cotovicus, Ioannes. 1619. *Itinerarium Hierosolymitanum et Syriacum* (Antverpiae).

Crane, E. 1975. *Honey: a comprehensive survey* (London).

Craven, R. K. 1821. *Tour through the southern provinces of the kingdom of Naples* (London).

Crescentius, Petrus de. 1548. *De omnibus agriculturae partibus . . .* (Basileae).

Crews, C. 1967. "One hundred medical recipes in Judeo-Spanish of ca. 1600." *Revue des études juives* 126: pp. 203–263.

Crowfoot, G. M. and L. Baldensperger. 1932. *From cedar to hyssop: a study in the folklore of plants in Palestine* (London).

Culpeper, Nicolas. 1932. *The English Physician and Complete Herbal* (ed. W. J. Ferrier, London).

Curtius Rufus, Quintus. 1821. *A history of Alexander the Great* (ed. and trans. P. Pratt, 2 v., 2nd ed., London).

Curzon, G. N. 1892. *Persia and the Persian question* (2 v., London).

Dale, Samuel. 1693. *Pharmacologia, seu manuductio ad Materiam Medicam* (Londini).

Dalechamps, Jacques. 1586–1587. *Historia generalis plantarum, in libros XVIII* (2 v., Lugduni).

Dalman, G. 1928–1942. *Arbeit und Sitte in Palästina* (7 v., Gütersloh).

Darwin, C. 1876. *The effects of cross and self fertilization in the vegetable kingdom* (London).

Dastur, J. F. 1962. *Useful plants of India and Pakistan* (Bombay).

Daubeny, C. 1865. *Essay on the trees and shrubs of the Ancients* (Oxford).

Davis, P. H. (ed.). 1965–1978. *Flora of Turkey* (6 v., Edinburgh).

Decaisne, J. 1834–1835. "Florula Sinaica. Énumération des plantes recueillies par M. Bové dans les deux Arabies, la Palestine, la Syrie et l'Égypte." *Annales des sciences naturelles* 2ième sér. 2: pp. 1–18, 239–270; 3: pp. 257–291.

—— 1847. "Remarks on a supposed heavenly manna." *Horticulturalist* 1, 7: pp. 313–315.

Deerr, N. F. 1949–1950. *The history of sugar* (2 v., London).

Deines, H. Von, H. Grapow and W. Westendorf. 1954–1973. *Uebersetzung der Medizinischen Texte: Grundriss Medizin der Alten Ägypter* (9 v. in 11, Berlin).

De La Hire (M.). 1709. "Observation botanique." *Histoire de l'Academie Royale des Sciences de Paris*, année 1708, 1709: p. 69.

Délile, A. R. 1812. "Mémoire sur les plantes qui croissent spontanément en Egypte." et "Florae Aegyptiacae." In: *Description de l'Égypte: Histoire Naturelle* (ed. E. F. Jomard, 2 v., 1809–1812, Paris) 2: pp. 1–10, 49–82, 145–320.

Depping, G. B. 1830. *Histoire du commerce entre le Levant et l'Europe, depuis les croisades jusqu'à la fondation des colonies d'Amérique* (2 v., Paris).

Deusingius, Antonius. 1659. *Dissertationes de manna, et saccharo* (Groningae).

Dey, K. L. 1896. *Indigenous drugs of India* (Calcutta).

Dickinson, J. C. 1969. "The eucalypt in the sierra of southern Peru." *Annals of the Association of American Geographers* 59, 2: pp. 294–307.

Diels, L. 1901. *Die Flora von Central-China* (Leipzig).

Dierbach, J. H. 1826. "Materia medica von Hindostan." *Geiger's Magazin für Pharmacie* 13: pp. 9–30.

Dillenius, J. J. 1732. *Hortus Elthamensis, seu plantarum rariorum quas in horto suo Elthami in Cantio coluit* (2 v., Londini).

Dillon, J. T. 1780. *Travels through Spain* (London).

Dinsmore, J. E. and G. Dalman. 1911. "Die Pflanzen Palästinas." *Zeitschrift des Deutschen Palästina-Vereins* 34: pp. 1–38, 147–172, 185–241.

Diodorus Siculus. 1958–1967. *Diodorus of Sicily* (ed. and trans. R. M. Geer, 12 v., Loeb Classics, London).

Dioscorides, Pedanius. 1829–1830. *Pedanii Dioscoridis Anazabei de Materia Medica Libri Quinque*. Medicorum Graecorum Opera Quae Exstant 25–26 (ed. C. G. Kühn, 2 v., Lipsiae).

——1952–1959. *La Materia Medica de Dioscórides* (traducida y commentada por D. Andrés de Laguna [1555], ed. C. E. Dubler, 6 v., Barcelona).

—— 1934. *The Greek Herbal of Dioscorides* (illustrated by a Byzantine A.D. 512, trans. J. Goodyer 1655, ed. R. W. T. Gunther, Oxford).

Djavanchir-Khoie, K. 1967. *Les chênes de l'Iran* (Montpellier).

Dobson, T. 1851. "On laap or lerp, the cup-like coverings of Psyllidae found on the leaves of certain Eucalypti." *Papers and Proceedings of the Royal Society of Van Dieman's Land* 1: pp. 235–241.

Dodonaeus, Rembertus. 1583. *Stirpium historiae pemptades sex sive libri XXX* (Antverpiae).

Don, G. 1825. *Prodromus Florae Nepalensis* (London).
—— 1831–1838. *History of dichladymeous plants* (4 v., London).
Dorostkar, H. 1974. *Contribution a l'étude des forêts du district Hyrcanien oriental* (Gembloux, Belgique).
Dorvault, F. L. M. 1844. *L'Officine ou répertoire général de pharmacie pratique* (Paris).
Dos Sanctos, Joaõ. 1905. "Collections out of the voyage and historie of Friar Joaõ dos Sanctos." In: Samuel Purchas *Hakluytus Posthumus or Purchas his Pilgrimes* (20 v., 1905–1907, Glasgow) 9: pp. 197–255.
Doughty, C. M. 1936. *Travels in Arabia Deserta* (2 v., London).
Dragendorff, G. 1898. *Die Heilpflanzen der Verschiedenen Völker und Zeiten* (Stuttgart).
Drury, H. 1873. *The useful plants of India* (2nd ed., Madras).
Drury, Robert. 1890. *Madagascar, or Robert Drury's journal during fifteen years captivity on that island* [London, 1729] (ed. S. P. Oliver, London).
Dubois-Aymé, J. M. J. 1809. *Notice sur le séjour des Hébraux en Egypt et sur leur fuite dans le désert* (Paris).
Duchesne, E. A. 1836. *Répertoire des plantes utiles et des plantes vénéneuses du globe* (Paris).
Ducloux, E. Herrero. 1928. "Nota sobre un maná indigena." *Revista de la facultad de ciencias quimicas* (Universidad Nacional de la Plata) 5, 1: pp. 73–76.
Ducros, M. A. H. 1930. *Essai sur le droguier populaire arabe de l'inspectorat des pharmacies du Caire*. Mémoires présentés à l'Institut d'Egypte, 15 (Le Caire).
Dufrené, H. 1887. *La Flore Sanscrite* (Paris).
Duhamel du Monceau, H. L. 1755. *Traité des arbres et arbustes* (2 v., Paris).
—— 1758. *La physique des arbres* (2 v., Paris).
Dümichen, J. 1867. *Altägyptische Tempelinschriften* (2 v., Leipzig).
Dupré, A. 1819. *Voyage en Perse* [1807–1809] (2 v., Paris).
Dutt, U. C. 1877. *The materia medica of the Hindus* (Calcutta).
Duveyrier, H. 1864. *Les Touareg du Nord* (Paris).
Dymock, W. 1885. *The vegetable materia medica of western India* (5 parts, London and Bombay).
—— 1890–1893. *Pharmaeographia Indica* (3 v., London).
Ebbell, B. (ed. and trans.) 1937. *The Papyrus Ebers* (Copenhagen and London).
Ebers, G. 1872. *Durch Gosen zum Sinai* (Leipzig).
—— 1875. *Papyros Ebers* (2 v., Leipzig).
Ebert, A. 1908. "Beiträge zur Kenntnis einiger seltener Mannasorten und verwandter Körper." *Zeitschrift des allgemeine österreichische Apotheker-Vereines* 46: pp. 427–429, 439–440, 447–450, 459–460, 467–470, 479–481, 491–492, 503–504, 515–516.
Ehrenberg, C. G. 1827a. "Ueber die Manna-Tamariske nebst allgemeinen Bemerkungen über die Tamariscineen." *Linnaea* 2: pp. 241–282.
—— 1827b. "Observations sur la famille des Tamariscinées, et sur la manne du tamarisque du mont Sinai." *Annales des Sciences Naturelles* (Paris) 12: pp. 68–78.
Ehrenberg, C. G. and F. W. Hemprich. 1900. *Symbolae Physicae: Botanica* [1820–1825] (Berolini).
Elenkin, A. 1901a. "Wanderflechten der Steppen und Wüsten." *Bulletin du Jardin Botanique de St. Petersbourg* 1: pp. 16–38, 52–72.
—— 1901b. "Lichenes Florae Rossiae et regionum confinium orientalium I." *Acta Horti Petropolitani* 19, 1: pp. 1–52.
Elphinstone, M. 1969. *An Account of the Kingdom of Caubul* [1815] (Graz).
Endlicher, S. L. 1841. *Enchiridion Botanicum* (Lipsiae, Viennae).
Engeström, Jacob Von. 1776. "Berättelse om Et Slags Socker funnit på *Granqvistar*." *Physiographisk Sällskapets Handlingar* (Stockholm) 1, 3: pp. 144–150.
Erman, A. and H. Grapow. 1926–1931. *Wörterbuch der aegyptischen Sprache* (5 v., Leipzig).
Errera, L. 1893. "Sur "pain du ciel" provenant de Diarbekir." *Bulletin de l'Académie Royale des Sciences, des Lettres et des Beaux-Arts de Belgique* 3ième sér. 26: pp. 83–91.
Eversmann, E. 1823. *Reise von Orenburg nach Bukhara* (Berlin).
—— 1831. "Lichenem Esculentum Pallasii: mit einem Nachtrage von Dr Fr L. von Esenbeck." *Nova Acta Physico-Medico, Academiae Caesareae Leopoldino-Carolinae* (Breslau und Bonn) 15, 2: pp. 351–362.
Exell, A. W., A. Fernandes, and H. Wild (eds.) 1960–1966. *Flora Zambesiaca* (London).
Eyre, E. J. 1845. *Journals of expeditions of discovery into Central Australia, and overland from Adelaide to King George's Sound, in the years 1840–1841* (2 v., London).
Fabri, Felix. 1892–1893. *The book of the wanderings of Felix Fabri* (ed. and trans. A. Stewart, 2 v., London).

130

Fabri, Ioannis Ernesti. 1776. "De Manna Ebraeorum Opuscula." In: I. I. Reiske and I. E. Fabri *Opuscula Medica ex Monimentis Arabum et Ebraeorum* (Halae), pp. 83–140.

Fahir İz and H. C. Hony. 1952. *An English-Turkish Dictionary* (Oxford).

Faldermann, F. 1837. *Fauna Entomologica Trans-Caucasica II* (Moscou).

Faurel, L., P. Ozenda, and G. Schotter. 1953. "Les lichens du Sahara algérien." *Research Council of Israel, Special Publications* 2: pp. 310–317.

Fazakerley, J. 1820. *Journey from Cairo to Mount Sinai, and return to Cairo* [1811]. In: R. Walpole (ed.) *Travels in various countries of the East* (London), pp. 362–391.

Fazelli, Tommaso. 1558. *De Rebus Siculis Decades Duae* (Panormi).

Fedtshenko, B. A. 1928. "Flora Rossiae Austro-Orientalis." *Acta Horti Petropolitani* 40: pp. 75–256.

Fée, M. A. and L. A. 1825. "Matière médicale de l'Indoustan." *Journal de Chimie Médicale, de Pharmacie et de Toxicologie* 1825: pp. 252–258, 290–297.

Ferrier, J. P. 1856. *Caravan journeys and wanderings in Persia, Afghanistan and Beloochistan* (ed. H. D. Seymour, trans. W. Jesse, London).

Fiore da Cropani, Giovanni. 1691–1743. *Della Calabria Illustrata* (2 v., Napoli).

Fischer, F. E. L. Von. 1812. *Catalogue du Jardin des Plantes de son excellence monsieur le comte Alexis de Razoumoffsky à Gorenki près de Moscou* (2nd ed., Moscou).

Fisher, E. E. 1945. "Manna formation in *Myoporum platycarpum*." *Journal of the Council for Scientific and Industrial Research* (Melbourne) 18: pp. 159–164.

Flacourt, Étienne de. 1658a. *Histoire de la grande isle de Madagascar* (Paris).

—— 1658b. *Dictionnaire de la langue de Madagascar* (Paris).

—— 1905. *Dictionnaire de la langue de Madagascar* (Paris).

Flagey, C. 1891–1892. "Lichenes algeriensis exsiccati." *Revue Mycologique* 13: pp. 83–87, 107–117; 14: pp. 70–79.

—— 1896–1897. *Catalogue des lichens*. In: J. A. Battandier and L. Trabut (eds.) *Flore d'Algérie* (2ième sér., Algér) 1.

Flückiger, F. A. 1868a. "Lerp Manna von *Eucalyptus dumosa*." *Jahresbericht über die Fortschritte der Pharmacognosie, Pharmacie und Toxicologie* 3: p. 124.

—— 1868b. "Ueber die Lerp-Manna." *G. C. Wittstein's Vierteljahrsschrift für praktische Pharmazie* 17: pp. 161–169.

—— 1871a. [On lerp manna]. *Yearbook of Pharmacy* 1871: pp. 188–190.

—— 1871b. "Ueber Stärke und Cellulose." *Archiv der Pharmazie* 196: pp. 7–29.

—— 1872. "Notiz über die Eichenmanna von Kurdistan." *Archiv der Pharmazie* 200: pp. 154–164.

—— 1883. *Pharmakognosie des Pflanzenreiches* (Berlin).

—— 1894. "Australische Manna." *Archiv der Pharmazie* 232: pp. 311–314.

Flückiger, F. A. and D. Hanbury. 1879. *Pharmacographia: a history of the principal drugs of vegetable origin met with in Great Britain and British India* (2nd ed., London).

Forbes, F. B., and W. B. Hemsley. 1886–1905. "Index Florae Sinensis: Ennumeration of all the plants known from China proper, Formosa, Hainan, Corea ... the island of Hong Kong." *Journal of the Linnean Society: Botany* 23, 26, 36.

Forbes, R. J. 1966. "Sugar and its substitutes in antiquity." In: *Studies in Ancient Technology* (2nd ed., Leiden) 5: pp. 80–111.

Ford, K. C. 1976. *Las hierbas de la gente: a study of Hispano-American medicinal plants* (University of Michigan Museum of Anthropology, paper 60, Ann Arbor).

Forskål, Petrus. 1775a. *Flora Aegyptiaco-Arabica, sive Descriptiones Plantarum* (Havniae).

—— 1775b. *Descriptiones Animalium .... adjuncta est Materia Medica Kahirina* (Havniae).

Fothergill, John. 1746. "Observations on the Manna Persicum." *Philosophical Transactions of the Royal Society of London* 43, 472: pp. 86–94.

Fraas, C. N. 1845. *Synopsis plantarum florae classicae* (München).

Franchet, A. 1883–1888. "Plantae Davidianae ex Sinarum Imperio." *Nouvelles Archives du Muséum d'Histoire Naturelle* (Paris) 2ième sér. 5–8, 10.

Fraser, J. B. 1826. *Narrative of a voyage to Khorasan* (London).

—— 1834. *An historical and descriptive account of Persia* (2nd ed., Edinburgh).

Frauenfeld, G. Von. 1859. "Ueber exotische Pflanzenauswüchse, erzeugt von Insecten." *Verhandlungen der zoologisch-botanischen Gesellschaft in Wein* 9: pp. 319–332.

Frederick, E. 1819. "Remarks on the substance called *gez* or *manna*, found in Persia and Armenia." *Transactions of the Literary Society of Bombay* 1: pp. 251–258.

Frescobaldi, Leonardo, Giorgio Gucci, and Simone Sigoli. 1948. *Visit to the Holy Places of Egypt, Sinai, Palestine and Syria in 1384* (trans. Fr. T. Bellorini and Fr. E. Hoade. Publications of the Stadium Biblicum Franciscanum 6, Jerusalem).

131

Fresenius, J. B. G. W. 1834. *Beiträge zur Flora von Aegypten und Arabien* (Frankfurt am Main).

Fryer, John. 1909–1915. *A New Account of East India and Persia, being nine years travels, 1672–1681* (ed. W. Crooke, Hakluyt Society, 2nd ser., 3 v. [XIX, XX, XXXIX], London).

Fuchsius, Leonardus. 1535. *Paradoxorum Medicinae Libri Tres* (Basileae).

Fürer ab Haimendorf, Christoph. 1621. *Itinerarium Aegypti, Arabiae, Palestinae, Syriae* .... [1565–1566] (Norimbergae).

Furlong, J. R. and L. E. Campbell. 1913. "A new variety of manna and a note on the melting point of dulcitol." *Proceedings of the Chemical Society* 29: p. 128.

Galenus, Claudius. 1530. *De Alimentorum Facultatibus Libri III* (Parisiis).

—— 1825–1830. *Opera Omnia.* In: C. G. Kühn (ed.) *Medicorum Graecorum* (20 v., 1821–1833, Lipsiae) 10, 11, 12, 17, 19.

Gamble, J. C. 1915–1936. *Flora of the Presidency of Madras* (3 v., London).

Garidel, P.-J. 1719. *Histoire de plantes qui naissent en Provence, et principalement aux environs d'Aix* (Aix et Paris).

Gast, M. 1968. *Alimentation des populations de l'Ahaggar.* Mémoires du Centre de Recherches Anthropologiques, Préhistoriques et Ethnographiques 8 (Paris).

Gaster, M. (ed. and trans.) 1899. *The Chronicles of Jeraḥmeel or The Hebrew Bible Historiale.* Oriental Translation Fund, new series, 4 (London).

Gauba, E. 1949–1953. "Botanische Reisen in der persischen Dattelregion." *Annalen des Naturhistorischen Museums in Wien* 57: pp. 42–52; 58: pp. 13–32; 59: pp. 119–134.

Geoffroy, Étienne François 1736. *A treatise of the fossil, vegetable and animal substances that are made use of in physick* (trans. G. Douglas, London).

—— 1741. *Tractatus de Materia Medica* (ed. E. Chardon de Courcelles, 3 v., Paris).

—— 1749. *A treatise on foreign vegetables ... chiefly taken from the Materia Medica of S. F. Geoffroy* [by R. Thicknesse] (London).

Georgius [Syncellus] 1652. *Chronographia* (Parisiis).

Gerarde, John. 1597. *The Herbal or Generall Historie of Plantes* (London).

Gerth Van Wijk, H. L. 1911. *A Dictionary of Plant Names* (The Hague).

Ghistele, Joos Van. 1557. *Tvoyage van Mher Joos Van Ghistele oft Anders* [1485] (Ghendt).

Gilliat-Smith, B. and W. B. Turril. 1930. "On the flora of the Nearer East: a contribution to our knowledge of the flora of Azerbaidjan, North Persia." *Kew Bulletin* 1930, 7: pp. 273–312; 8: pp. 375–398; 9: pp. 427–463; 10: pp. 480–493.

Gmelin, Samuel Gottlieb. 1770–1774. *Reise durch Russland* (2 v. in 1, St. Petersburg).

—— 1774. "Reise durch das nordliche Persien." In: *Reise durch Russland* (4 v. in 3, 1774–1784, St. Petersburg) 3.

Goebel, (Fr.) 1830. "Chemische Untersuchung einer in Persien herabgeregneten Substanz, der *Parmelia esculenta.*" *Jahrbuch der Chemie und Physik* 3: pp. 393–399.

Goetz, G. (ed.) 1888–1923. *Corpus Glossariorum Latinorum* (7 v., Lipsiae).

González de Clavijo, Ruy. 1859. *Narrative of the Embassy of Ruy González de Clavijo to the Court of Timour at Samarcand, A.D. 1403–1406* (ed. and trans. C. R. Markham, Hakluyt Society, XXII, London).

—— 1928. *Embassy to Tamerlane* [1403–1406] (ed. and trans. G. Le Strange, London).

Göppert, H. R. 1831. "Ueber die sogenannten Getreide- und Schwefel-Regen." *Poggendorff's Annalen der Physik und Chemie* 21: pp. 550–578.

Gorski, (Prof.) 1912. In: Reports, mainly reprinted from the *Spenersche Zeitung* and the *Vossische Zeitung, Sitzungsberichte der Gesellschaft naturforschender Freunde zu Berlin* [1839–1859] (Berlin), pp. 76–77.

Grandidier, A. 1917. *Histoire naturelle, physique, et politique de Madagascar* (36 v., 1882–1928, Paris) 4, 3.

Grassé, P.-P. 1949–1951. *Traité de Zoologie* (2 v. [IX, X], Paris).

Green, T. 1820. *The Universal Herbal* (2 v., Liverpool).

Griffith, W. 1847. *Journals of travels in Assam, Burma, Bootan, Afghanistan and the neighbouring countries* (Calcutta).

Gubb, A. S. 1913. *La flore saharienne* (Alger).

Gueldenstadt, J. A. 1787–1791. *Reisen durch Russland und in Kaukasischen Gebirge in den Jahren 1768–1773,* heraus gegeben von P. S. Pallas (St. Petersburg).

Guest, E. 1932. *Notes on trees and shrubs for Lower Iraq* (Baghdad).

—— 1933. *Notes on plants and plant products with their colloquial names in Iraq* (Baghdad).

Guest, E. and G. C. Townsend (eds.) 1966–1974. *Flora of Iraq* (3 v., Baghdad).

Guibourt, N. J. B. G. 1849–1851. *Histoire naturelle des drogues simples* (4 v., Paris).

—— 1858a. "Notice sur une matière pharmaceutique nommée le tréhala, produite par un insecte de la famille des Charançons."' *Comptes Rendus Hebdomadaires des Séances de l'Academie des Sciences* (Paris) 46: pp. 1213–1217.

—— 1858b. "Notice sur le tréhala." *Journal de pharmacie et de chimie* 3ième sér. 34: pp. 81–87.

Guigues, P. 1905. "Les noms arabes dans Sérapion, *Liber de Simplici Medicina.*" *Journal Asiatique* 10ième sér. 5: pp. 473–546; 6: pp. 49–112.

Guillaumin, A. 1946. *Les plantes cultivées, histoire, économie* (Paris).

Gunn, 1841. "Remarks on the indigenous vegetable productions of Tasmania available to man." *Tasmanian Journal of Natural Science* 1: pp. 35–52.

Guyon, J. L. G. 1852. *Voyage d'Alger aux Ziban, l'anciénne Zebe, en 1847* (Alger).

Hackel, E. 1887. "Gramineae." In: A. Engler and K. Prantl (eds.) *Die natürlichen Pflanzenfamilien* (4 v., 1887–1915, Leipzig) 2, 2.

Haidinger, W. 1864. "Ein Mannaregen bei Charput in Kleinasien im März 1864." *Anzeiger der Kaiserlichen Akademie der Wissenshaften in Wien: Mathematisch-Naturwissenschaftliche Classe* 1, 18: pp. 129–130.

—— 1865. "Ein Mannaregen bei Karput in Klein-Asien im März 1864." *Sitzungsberichte der Mathematisch-Naturwissenschaftlichen Classe der Kaiserlichen Akademie der Wissenschaften in Wien* 50, 2: pp. 170–177.

Hakluyt, Richard. 1903–1905. *The Principal Navigations, Voyages, Traffiques and Discoveries of the English Nation* (12 v., Glasgow).

Hale, M. E. 1967. *The Biology of Lichens* (London).

Hallé, (M.) 1787. "Agul." and "Alhagi." In: *L'Encyclopédie méthodique* (190 v., 1782–1830, Liège and Paris): *Médicine* 1: pp. 397–399, 673–674.

Haller, Albert Von. 1768. *Historia Stirpium Indigenarum Helvetiae* (2 v., Bernae).

Hamarneh, S. K. 1973. *Origins of Pharmacy and Therapy in the Near East* (Tokyo).

Hamarneh, S. K. and G. Sonnedecker. 1963. *A pharmaceutical view of Abulcasis al-Zahrāwī in Moorish Spain, with special reference to the "Adhān"* (Leiden).

Ḥamarneh-Allāh Mustaufī. 1919. *The Geographical Part of the Nuzat-al-Qulūb, composed by Ḥamd-Allāh Mustaufī of Qazwīn in 740* [1340] (ed. and trans. G. Le Strange, London).

Hamelius, Ludovicus Sebaldus. 1723. *Dissertatio philologico-theologica de urna mannae ejusque mysterio* (Herbornae Nassaviorum).

Hampe, E. 1848. "Ueber *Lichen esculentus.*" *Botanische Zeitung* 6, 52: cols. 889–891.

Hanbury, D. 1859. "Note on two insect products from Persia." *Journal of the Proceedings of the Linnean Society of London* 3: pp. 178–183.

—— 1862–1863. "Minor notes on the Materia Medica of the International Exhibition." *Pharmaceutical Journal and Transactions* 2nd ser. 4: pp. 107–111.

—— 1870. "Historical notes on manna." *Pharmaceutical Journal and Transactions* 2nd ser. 11: pp. 326–331.

—— 1876. *Science papers, chiefly pharmacological and botanical* (London).

Hanoteau, A. and A. Letourneux. 1872–1873. *La Kabylie et les coutumes kabyles* (3 v., Paris).

Hardwick, T. 1822. "Description of a substance called *gez* or *manna*, and of the insect producing it." *Asiatic Researches* 14: pp. 182–186.

Hare, H. A., C. Caspari, and H. H. Rusby. 1905. *The National Standard Dispensatory* (London).

Harff, Arnold Von. 1946. *The pilgrimage of Arnold von Harff, Knight* (ed. and trans. M. Letts, Hakluyt Society, 2nd ser., XCIV, London).

Harlan, J. 1939. *Central Asia* [1823–1841] (London).

Harris, John (ed.) 1764. *Navigantum Itinerantium Bibliotheca* (2 v., revised ed., London).

Harrison, R. K. 1966. *Healing herbs of the Bible* (Leiden).

Harrison, S. G. 1951. "Manna and its sources." *Kew Bulletin* 1951: pp. 407–417.

Hart, H. C. 1885. "Report on the botany of Sinai and south Palestine." *Transactions of the Royal Irish Academy* 28: pp. 373–452.

—— 1891. *Some account of the fauna and flora of Sinai, Petra, and Wady Arabah* (London).

Hartwich, C. and G. Håkanson. 1905. "Über *Glyceria fluitans*, ein fast vergessenes einheimisches Getreide." *Zeitschrift für Untersuchung der Nahrungs- und Genussmittel* 10: pp. 473–478.

Hasselquist, Frederik. 1766. *Voyages and travels in the Levant* [1749–1752] (London).

Haupt, P. 1922. "Manna, nectar, and ambrosia." *Proceedings of the American Philosophical Society* 61: pp. 227–236.

Haussknecht, A. 1870. "Ueber Manna-Sorten des Orients." *Archiv der Pharmazie* 192: pp. 244–251.

Hawks, Henry. 1904. *A Relation of the Commodities of Nova Hispania* [1572]. In: Richard

Hakluyt, *The Principal Navigations, Voyages, Traffiques and Discoveries of the English Nation* (12 v., 1903–1905, Glasgow) 9: pp. 378–397.

Haynald, L. 1894. *Des plantes qui fournissent les gommes et les résines mentionnés dans les livres saints* (Musée national hongrais, Budapest).

Hegi, G. 1912. *Illustrierte Flora von Mittel-Europa* (7 v. in 13, 1906–1931, München) 3.

Heldreich, T. Von. 1862. *Die Nutzpflanzen Greichenlands* (Athen und Leipzig).

Helms, R. 1896. "Anthropology of the Elder Exploring Expedition." *Transactions and Proceedings of the Royal Society of South Australia* 16, 3: pp. 238–332.

Hemsley, W. B. 1879–1888. *Biologia Centrali-Americana* (4 v., London).

Henneguy, F. 1883. *Les lichens utiles* (Paris).

Henniker, F. 1823. *Notes during a visit to Egypt, Nubia, the Oasis, Mount Sinai and Jerusalem* (London).

Henry, A. 1924. "Manna of larch and Douglas fir, melezitise and lethal honey." *Pharmaceutical Journal and Transactions* 4th ser. 58: pp. 387–390.

—— 1926. "Manna on larch trees." *Pharmaceutical Journal and Transactions* 4th ser. 63: pp. 300–301.

Herbelot, Bathélemy d'. 1697. *Bibliothèque orientale ou Dictionnaire universel* (Paris).

—— —— 1777–1779. *Bibliothèque orientale ou Dictionnaire universel* (4 v., La Haye).

Herbert, Thomas. 1928. *Travels in Persia* [1627–1629] (ed. W. Foster, London).

Hermann, F. 1956. *Flora von Nord- und Mitteleuropa* (Stuttgart).

Hernández, Francisco. 1959. *Historia natural de Nueva España* (trad. J. Rojo Navarro, 2 v. [of 3 v. of *Obras Completas*] México).

Hernández-Pacheco, E. 1949. *El-Sáhara español* (Madrid).

Herodotus. 1963–1969. *Herodotus* (ed. and trans. A. D. Godley, 4 v., Loeb Classics, London).

Herrera [y Tordesillas], Antonio de. 1934–1956. *Historia general de los hechos de los castellanos* [ca. 1600] (ed. A. Ballesteros-Beretta, 15 v., Madrid).

Herzfeld, E. E. 1968. *The Persian Empire: studies in the geography and ethnography of the ancient Near East* (ed. G. Walser, Weisbaden).

Hesiod. 1967. *The Homeric Hymns and Homerica* (ed. and trans. H. G. Evelyn-White, Loeb Classics, London).

Heyd, W. 1886. *Histoire du commerce du Levant au Moyen-Age* [*Geschichte des Levantehandels im Mittelalter*, 2 v., Stuttgart, 1879] (2 v., Leipzig).

Hillscher, S.-M. 1747. *Prolusio de Gramine Manna Dicta* (Jenae).

Hilton-Simpson, M. W. 1922. *Arab medicine and surgery: a study of the healing art in Algeria* (London).

Hippocrates. 1825–1827. *Magni Hippocratis Opera Omnia.* In: C. G. Kühn (ed.) *Medicorum Graecorum* (26 v., 1821–1833, Lipsiae) 21, 22, 23.

Hirth, F. and W. W. Rockhill (ed.) 1911. *Chau Ju-kua: His Work on the Chinese and Arab Trade in the XIIth and XIIIth Centuries Entitled Chu-fan-chi* (St. Petersburg).

Hodīvālā, S. H. 1939. *Studies in Indo-Muslim History* (Bombay).

Hoffmann, Friedrich. 1748–1754. *Opera Omnia* (6 v. in 3, Genevae).

Hogg, J. 1849. "Remarks on the manna of the Israelites." *Transactions of the Royal Society of Literature* 2nd ser. 3: pp. 183–236.

—— 1864. "On the recent fall of manna lichen." *The Reader* 1864: p. 205.

Holmes, E. M. 1920. "The Manna of the Scripture." *Journal of the Philadelphia College of Pharmacy* [*American Journal of Pharmacy*] 92: pp. 174–179.

Honigberger, J. M. 1852. *Thirty-five years in the East* (2 v., London).

Hooker, J. D. 1854. *Himalayan Journals* (2 v., London).

—— 1872–1897. *The Flora of British India* (7 v., London).

Hooker, W. J. 1843. [*Eucalyptus splachnicarpon*] *Curtis's Botanical Magazine*, new ser. 16: no. 4036.

Hooper, D. 1900. "Bamboo manna." *The Agricultural Ledger* 7, 17: pp. 185–189.

—— 1909. "Tamarisk manna." *Journal and Proceedings of the Asiatic Society of Bengal*, new ser. 5: pp. 31–36.

—— 1913. "Saracolla." *Journal and Proceedings of the Asiatic Society of Bengal*, new ser. 9: pp. 177–181.

—— 1929. "On Chinese medicine: drugs of Chinese pharmacies in Malaya." *The Gardens' Bulletin* (Straits Settlements) 6, 1–5: pp. 1–163.

—— 1931. "Persian drugs." *Kew Bulletin* 1931, no. 6: pp. 299–352.

Hooper, D. and H. Field. 1937. *Useful plants and drugs of Iran and Iraq* (Field Museum of Natural History, Botanical Series 9, 3, Chicago).

Horace. 1968. *The Odes and Epodes* (ed. and trans. C. E. Bennett, Loeb Classics, London).

134

Houel, J. P. 1782–1787. *Voyage pittoresque des Isles de Sicile, de Malte, et de Lipari* (4 v., Paris).

Houghton, W. 1864. "On the manna-lichen and the manna of scripture." *The Reader* 1864: p. 238.

Howes, F. N. 1949. *Vegetable gums and resins* (Waltham, Mass.).

Howitt, A. W. 1890. "The eucalypts of Gippsland." *Transactions of the Royal Society of Victoria* 2: pp. 83–113.

Huber, C. 1891. *Journal d'un voyage en Arabie, 1883–1884* (Paris).

Hudson, C. S. and S. F. Sherwood. 1918. "The occurrence of melesitose in a manna from the Douglas fir." *Journal of the American Chemical Society* 40: pp. 1456–1460.

Hue, A. M. 1891. "Lichenes exoticos." *Nouvelles archives du Muséum d'Histoire Naturelle* (Paris) 3ième sér. 3: pp. 33–192.

Hunter, A. E. 1969. "Down the treacle mine." [correspondence on the sweet exudate of mineralised deposits of *Fraxinus ornus*] *Pharmaceutical Journal* 202: p. 218.

Ibn al-'Awwām, Abū Zakariya. 1864–1867. *Le livre de l'agriculture* (trad. J. J. Clément-Mullet, 2 v., Paris).

Ibn al-Baiṭār, Abū Muḥammad. 1877–1883. *Traité des simples* (ed. et trad. L. Leclerc, Notices et Extraits des Manuscrits de la Bibliothèque National, 3 v. [XXIII, XXV, XXVI], Paris).

Ibn al-Balkhī. 1912. *Description of the province of Fars in Persia at the beginning of the twelfth century* (ed. and trans. G. Le Strange, London [reprinted from the *Journal of the Royal Asiatic Society* 1912: pp. 1–30, 311–339, 865–889]).

Ibn Buṭlān, Abū-l-Ḥasan. 1531. *Tacuinum sanitatis* (Argentorati).

Ibn Kaysān, Abū Sahl. 1953. In: P. Sbath and C. D. Avierinos (ed. et trad.) *Deux Traités Médicaux* (Le Caire).

Ibn Māsawaih, Abū Zakariya Yūḥanna. 1502. *De Consolatione Medicinarum* (Venetiis).

—— 1581. *Medici Clarissimi Opera* (Venetiis).

Ibn Milad, Aḥmad. 1933. *L'Ecole médicale de Kairouan aux Xe. et XIe. siècles* (Paris).

Ibn Sarābī. 1531. *De Simplicibus Medicinis Opus* (Argentorati).

Ibn Sīna [Avicenna], Abū 'Ali al-Ḥusain. 1608. *Avicennae Arabum Medicorum Principis* (2 v., Venetiis).

Irving, F. R. 1957. "Wild and emergency foods of Australian and Tasmanian aborigines." *Oceania* 28, 2: pp. 113–142.

Irwin (Lt.) 1839–1840. "Memoir on the climate, soil, produce and husbandry of Afghanistan and the neighbouring countries." *Journal of the Asiatic Society of Bengal* 8, 2: pp. 745–776, 779–804, 869–900, 1005–1015; 9, 1: pp. 33–64, 189–197.

'Īsā, Aḥmad. 1930. *Dictionnaire des noms des plantes en latin, français, anglais, et arabe* (Le Caire).

James, C. 1872. *Les Hébreux dans l'Isthme de Suez* (Paris).

Jandrier, E. 1893. "Sur le miellée du platane." *Comptes Rendus Hebdomadaires des Séances de l'Académie des Sciences* (Paris) 117: p. 498.

Janssen, J. 1879. "Fraxinus cultivation and manna production." *Pharmaceutical Journal and Transactions* 3rd ser. 10: p. 407.

Jaubert, H. F. and E. Spach. 1842–1857. *Illustrationes Plantarum Orientalium* (5 v., Paris).

Jekel, H. 1849. *C. J. Schoenherr Genera et Species Curculionidum Catalogus ab H. Jekel* (Paris).

Johnson, C. P. 1862. *Useful Plants of Great Britain* (London).

Johnston, J. F. W. 1843. "On the sugar of the Eucalyptus." *London, Edinburgh and Dublin Philosophical Magazine* 23: pp. 14–16.

Johnston, T. H. and J. B. Cleland. 1942. "Aboriginal names and uses of plants in the Ooldea region of South Australia." *Transactions of the Royal Society of South Australia* 66: pp. 93–103.

Johnstone, John. 1662. *Dendrographias, sive Historia Naturalis de Arboribus et Fructicibus* (Francofurti ad Moenum).

Jordanus (Fr.) 1863. *Mirabilia Descripta: the wonders of the East by Friar Jordanus* (ed. and trans. H. Yule, Hakluyt Society, XXXI, London).

Joret, C. 1897–1904. *Les plantes dans l'Antiquité et au Moyen Age* (2 v., Paris).

Josephus, Flavius. 1967–1969. *Jewish Antiquities* (ed. and trans. H. St. J. Thackeray, 6 v. [4 to 9 of complete works], Loeb Classics, London).

Jourdan, A. J. L. 1828. *Pharmacopée universelle* (2 v., Paris).

Juel, O. 1913. "Ett *Manna-Regn* I Botaniska Träd-Gården I Upsala." *Svensk Botanisk Tidskrift* 7, 2: pp. 189–195.

Kaiser, A. 1922. "Die Sinaiwüste." *Mitteilungen der Thurgauischen Naturforschenden Gesellschaft* 24: pp. 3–106.

135

—— 1924. "Der heutige Stand der Mannafrage." *Mitteilungen der Thurgauischen Naturforschenden Gesellschaft* 25: pp. 99–105.

Kämpfer, E. 1712. *Amoenitatum exoticarum politico-physico-medicarum fasciculi V* (Lemgoviae).

Keimer, L. 1924. *Die gartenpflanzen im alten Aegypten* (Hamburg und Berlin).

—— 1943. "Note sur le nom Égyptien du Jububier d'Égypte (*Zizyphus spina christi* Willd.)." *Annales du Service des Antiquités de l'Égypte* 42: pp. 279–281.

Keller, B. A. 1926. "Die Vegetation auf den Salzböden der russischen Halbwüsten und Wüsten." *Zeitschrift für Botanik* 18: pp. 113–137.

Keller, B. A. and K. K. Shaparenko. 1933. "Materialy k sistematiko-ekologicheskoi monografii roda *Alhagi* Tourn. ex Adans.". *Sovetskaya Botanika* 3–4: pp. 150–185.

Kiepert –. 1868. "Mittheilungen von C. Haussknecht's botanische Reisen in Kurdistan und Persien." *Zeitschrift der Gesellschaft für Erdkunde zu Berlin* 3: pp. 464–473.

Killian, C. 1947. *Biologie végétale au Fezzân* (Institut de Recherches sahariennes de l'Université d'Alger, Paris).

Kinneir, J. M. 1813. *A Geographical Memoir of the Persian Empire* (London).

Kircher, Athanasius. 1652–1654. *Oedipus Aegyptiaci* (3 v., Romae).

Kirtikar, K. R. and B. D. Basu. 1918. *Indian Medicinal Plants* (2 v., Allahabad).

Kitamura, S. 1960. *Flora of Afghanistan* (Kyoto).

Klein, G. 1931–1932. *Handbuch der Pflanzenanalyse* (6 v., Wien).

Klement, O. 1965. "Flechten aus der Mongolischen Volksrepublik." *Feddes Repertorium* 72: pp. 98–123.

Koch, K. 1846–1847. *Wanderungen in Oriente, 1843–1844* (2 v., Weimar).

Köhler, E. 1890. *Medizinal-Pflanzen* (ed. G. Papst, 4 v., Gera-Untermhaus).

Kolb, G. 1892. "Manna der Natur und der Bibel." *Natur und Offenbarung* 38: pp. 1–13.

Komarov, V. L. (ed.) 1934–1962. *Flora S.S.S.R.* (31 v., 1934–1964, Mosqua, Leningrad) 2, 18, 27.

Konrad von Megenberg. 1897. *Das Buch der Natur* (ed. H. Schulz, Greifswald).

Kotschy, T. 1862. *Die Eichen Europas und des Orients* (Wien und Olmüz).

Krempelhuber, A. Von. 1867. "*Lichen esculentus* Pall., ursprünglich eine steinbewohnende Flechte." *Verhandlungen der Kaiserlich-Königlichen zoologisch-botanischen Gesellschaft in Wien* 17: pp. 599–606.

Krünitz, J. G. 1808–1828. *Oekonomisch-tecknologische Enzyklopädie* (242 v., 1773–1858, Berlin), 83, 149.

Küchler, F. 1904. *Beiträge zur Kenntnis der assyrisch-babylonischen Medizin* (Assyriologische Bibliothek 18, Leipzig).

Labat, J.-B. 1730. *Voyages en Espagne et en Italie* (8 v., Paris).

—— 1742. *Nouveau voyage aux Isles de l'Amérique* [1693–1705] (8 v., Paris).

Labillardière, J. J. 1804–1806. *Novae Hollandiae Plantarum Specimen* (2 v., Parisiis).

Laborde, L. 1841. *Commentaire géographique sur l'Exode et les Nombres* (Paris et Leipzig).

Lacour, M. E. 1880. "Analyse chimique du *Lichen esculentus* (manne du désert ou manne des Hébreux)." *Répertoire de Pharmacie et Journal de Chimie médicale*, nouvelle série 8: pp. 449–453.

Laguna y Villaneuva, M. 1883–1890. *Flore Forestal Española* (4 v., Madrid).

Lambton, A. K. S. 1953. *Landlord and Peasant in Persia* (Oxford).

Landerer, X. 1842. "Ueber das Himmelsbrod der Hebräer vom Libanon." *Repertorium für die Pharmacie* (Nürnberg) Zweiter Reihe, 27: pp. 371–372.

—— 1854. "On the varieties of manna not produced by the ash." *Pharmaceutical Journal and Transactions* 13: pp. 411–412.

Landsberger, B. 1967. *The date palm and its by-products according to the cuneiform sources* (Archiv für Orientforschung, Beiheft 71, Graz).

Langdon, S. H. 1931. *The mythology of all races: V-Semitic* (Boston).

Langkavel, B. 1964. *Botanik der späteren Griechen* [1st ed. 1866, Berlin] (Amsterdam).

Langlois (M.) 1843a. "Chemische Untersuchung einer auf den Blättern der Linde gesammelten zuckerigen Substanz." *Journal für praktische Chemie* 29: pp. 444–447.

—— 1843b. "Examen chimique d'une matière sucrée recueillie sur les feuilles du tilleul." *Annales de Chimie et de Physique* 3ième sér. 7: pp. 348–351.

Lansdell, H. 1885. *Russian Central Asia* (2 v., London and New York).

Lapide, Cornelius a. 1866. *Commentaria in Scripturam Sacram* (ed. A. Crampon, 24 v., 1866–1878, Paris) 1.

Laufer, B. 1919. *Sino-Iranica* (Chicago).

Lawrence, R. 1968. *Aboriginal Habitat and Economy* (Department of Geography, Australian

National University, Canberra).

Layard, A. H. 1887. *Early adventures in Persia, Susiana and Babylonia* (2 v., London).

Lázaro é Ibiza, B. 1906–1907. *Compendio de la Flora Española y estudio especial de las plantas criptógamas y fanerógamas, indigenas y exóticos, que tienen aplicaciones en medicina* (2 v., 2nd ed., Madrid).

Leclerc, L. 1876. *Histoire de la Medicine Arabe* (2 v., Paris).

Le Coq, A. Von. 1911. *Sprichwörter und Lieder aus der Gegend von Turfan* (Baessler Archiv 1, Leipzig).

Ledebour, Carolus Fridericus. 1829–1833. *Flora Altaica* (4 v., Berolini).

—— 1842–1853. *Flora Rossica* (4 v., Stuttgartiae).

Ledel, Johann Samuel. 1733. *Succincta Mannae Excorticatio Betrachtung des Schwadens* (Sorau).

L'Héritier de Brutelle, C. L. 1784. *Stirpes Novae* (2 v., Parisiis).

Lehmann, A. 1852. *Reise nach Buchara und Samarkand* [1841–1842] Beiträge zur Kenntniss des Russischen Reiches 17 (St. Petersburg).

Le Huen, Nicole. 1488. *Les Saintes Pérégrinations de Jérusalem et des lieux prochains du mont Sinai* (Lyon).

Lémery, Nicolas. 1699. *Traité universel des drogues simples* (Paris).

Lengerke, Cäsar Von. 1844. *Kenáan: Volks- und Religionsgeschichte Israel's* (Königsberg).

Lenz, H. O. 1859. *Botanik der alten Griechen und Römer* (Gotha).

Leo Africanus. 1896. *The History and Description of Africa* . . . ., *done into English in the year 1600 by John Pory* (ed. R. Brown, Hakluyt Society, 3 v. [XCII, XCIII, XCIV], London).

Lepsius, K. R. 1846. *A Tour from Thebes to the peninsula of Sinai* (trans. C. H. Cottrell, London).

Le Strange, G. 1890. *Palestine under the Moslems: a description of Syria and the Holy Land* [A.D. 650 to 1500] . . . *from the works of the medieval Arab geographers* (London).

Leunis, J. 1844–1853. *Synopsis der drei Naturreiche* (3 v., Hannover).

Léveillé, J.-H. 1842. *Observations médicales et énumerations des plantes*. In: Anatole de Demidoff *Voyage dans la Russie méridionale* (4 v., 1840–1842, Paris) 2: pp. 41–242.

Levey, M. 1959. *Chemistry and Chemical Technology in Ancient Mesopotamia* (Amsterdam).

—— 1963. "Some facets of medieval Arabic pharmacology." *Transactions and Studies of the College of Physicians of Philadelphia* 30: pp. 157–162.

—— 1966. *The Medical Formulary or Aqrābādhīn of al-Kindī* (Madison).

—— 1973. *Early Arabic pharmacology* (Leiden).

Levshin, A. 1840. *Description des hordes et des steppes des Kirghiz-Kazaks* (trad. F. de Pigny, Paris).

Lindley, J. 1840. "Miscellaneous notices, nos. 72–74." *Edward's Botanical Register* 26: pp. 40–41.

—— 1853. *The Vegetable Kingdom* (3rd ed., London).

Lindley, J. and T. Moore (eds.) 1874. *The Treasury of Botany* (2 v., London).

Lindsay, A. W. C. 1838. *Letters on Egypt, Edom, and the Holy Land* (2 v., London).

Link, H. F. 1847. ["*Lecanora esculenta* und *L. affinis.*"] *Archiv für Naturgeschichte* 13, 2: pp. 247–248.

—— 1848. "Ueber *Lichen esculentus*, seine Reise nach Corsica u. a. Briefliche Mittheilung." *Botanische Zeitung* 6, 38: cols. 665–670.

—— 1849. "Ueber *Lichen Jussufii.*" *Botanisch Zeitung* 7, 41: cols. 729–731.

Linnaeus, Carolus. 1749. *Materia Medica* (Amstelaedami).

—— 1786. *Amoenitates Academicae* (8 v., Lugduni Batavorum).

—— 1938. *The Critica Botanica of Linnaeus* [1737] (ed. and trans. A. Hort, Ray Society, CXXIV, London).

—— 1957–1959. *Species Plantarum* [facs. of 1st ed. of 1753] (Ray Society, 2 v. [CXL, CXLII], London).

Linschoten, Jan Huygen Van. 1610. *Histoire de la Navigation de Iean Hugues de Linscot, avec annotation de Bernard Paludanus* (Amsterdam).

—— 1885. *Voyage to the East Indies* [1583–1592], *from the old English translation of 1958* (ed. A. C. Burnell and P. A. Tiele, Hakluyt Society, 2 v. [LXX, LXXI], London).

Lippmann, E. O. Von. 1890. *Geschichte des Zuckers* (Leipzig).

Littré, E. 1956–1958. *Dictionnaire de la Langue Française* (7 v., édition intégrale, Paris).

Littré, E. and C. Robin. 1865. *Dictionnaire de médecine, de chirurgie, de pharmacie* . . . . (Paris).

Lloyd, J. U. 1897. "The California manna." *American Journal of Pharmacy* 69: pp. 329–338.

Lobel, Matthiae de. 1576. *Plantarum seu stirpium historia* [part 2: *Stirpium adversa*, 1570–1571] (Antverpiae).

137

—— 1591. *Icones Stirpium* (Antverpiae).

Loiseleur-Deslongchamps, J. L. A. 1819. *Manuel des plantes usuelles indigènes* (2 parts, Paris).

—— 1837. *Histoire du cèdre du Liban* (Paris).

Loret, V. 1887. *La Flore pharaonique, d'après les documents hiéroglyphiques* (Paris).

Lovell, Robert. 1665. *Enchiridion Botanicum, or a Compleat Herball* (Oxford).

Löw, I. 1881. *Aramaeische Pflanzennamen* (Leipzig).

—— 1967. *Die Flora der Juden* [repr. of the 1924–1934 ed.] (4 v., Wien und Leipzig).

Luca, L. de. 1863. "On the formation of the fatty matter and mannite in the olive." *Pharmaceutical Journal and Transactions* 2nd ser. 4: p. 473.

Ludolphus de Suchen. 1895. *Description of the Holy Land and of the way thither* [1350] (ed. and trans. A. Stewart, London).

Ludwig, H. 1866. "Ein Mannaregen bei Karput in Kleinasien im März 1864." *Archiv der Pharmazie* 177: pp. 284–288.

—— 1870. "Ueber die Bestandtheile einiger Mannasorten des Orients." *Archiv der Pharmazie* 193: pp. 32–52.

Luerssen, C. 1879–1882. *Medicinisch-pharmaceutische Botanik* (2 v., Leipzig).

Lynch, H. F. B. 1901. *Armenia: Travels and Studies* (2 v., London).

Macgregor, C. M. (ed.) 1873–1875. *Gazetteer of Central Asia* (7 v., Calcutta).

Macpherson, J. 1939. "The eucalyptus in the daily life and medical practice of the Australian aborigines." *Mankind* 2, 6: pp. 175–180.

Madden, E. 1850. *Observations on Himalayan Coniferae* (Calcutta).

Maffei [Volaterranus], Raffaello. 1559. *Commentatiorum urbanorum, XXXVIII libri* [Paris, 1515] (Basiliae).

Maiden, J. H. 1889a. *Useful native plants of Australia* (London).

—— 1889b. "Australian indigenous plants providing human foods and food adjuncts." *Proceedings of the Linnean Society of New South Wales* 2nd ser. 3: pp. 481–556.

—— 1892. "Vegetable exudations." *Transactions and Proceedings of the Royal Society of South Australia* 16: pp. 1–9.

—— 1909–1933. *A critical revision of the genus Eucalyptus* (8 v., Sydney).

—— 1920. "Australian manna." In: *Forest Flora of New South Wales* (Sydney) Part LXIII: pp. 101–111.

Maimonides [Moses ben Maimon]. 1940. *L'explication des noms de drogues: un glossaire de matière médicale composé par Maimonides* (ed. and trans. M. Meyerhof, Mémoires présentés à l'Institut d'Égypte 41, Le Caire).

Maire, R. and T. Monod. 1950. *Études sur la flore et la végétation du Tibesti* (Paris).

Malcolm, J. 1815. *The History of Persia* (2 v., London).

Malina, B. J. 1968. *The Palestinian manna tradition* (Institutum Iudaicum, Tübingen).

Mandelseo, John Albert de. 1669. *Travels from Persia into the East Indies* [1633–1639] (trans. John Davies, London).

Manucci, Niccolao. 1907–1908. *Storia Do Mogor or Mogul India 1653–1708, by Niccolao Manucci, Venetian* (ed. and trans. W. Irvine, 4 v., London).

Marcorelle, J.-F. 1760. "Extrait d'une lettre sur une espèce de manne qui croit sur des saules et des frênes aux environs de Carcassonne." *Mémoires de Mathematique et de Physique* (Académie Royale des Sciences, Paris) 3: pp. 501–503.

Marilaun, A. Kerner Von. 1894–1895. *The natural history of plants* (ed. and trans. F. W. Oliver, 2 v., London).

—— 1896. "Über das Vorkommen der Manna-Flechte (*L. esculenta*) in Griechenland." *Anzeiger der Kaiserlichen Akademie der Wissenschaften in Wien. Mathematisch-Naturwissenschaftliche Classe* 33, 5: pp. 35–37.

Markownikoff, V. 1885. "Turkestan manna [abstract]." *Journal of the Chemical Society* 48, 1: p. 943.

Martius, H. Von. 1852. "Ueber die sogenannte Manna von Sidi Ghafi Batal in Kleinasien." *Gelehrten Anzeigen* (München) 34: cols. 20–21.

Masson, Charles (pseudonym). 1842. *Narrative of various journeys in Balochistan, Afghanistan and the Panjab* (3 v., London).

—— 1843. *Narrative of a journey to Kalât* (London).

Masson, P. 1896. *Histoire du commerce français dans le Levant au XVIIe. siècle* (Paris).

Mas y Guindal, A. 1953. "En el centenario de la muerte de Cristóbal Velez." *Boletín de la Sociedad Española de Historia de la Farmacia* 14: pp. 78–85.

Matthiolus, Pietro Andrea. 1544. *Di P. Dioscoride libri cinque della historia* (Venetia).

—— 1558. *Commentarii secundo aucti, in libros sex Pedacii Dioscoridis Anazarbei de Medica Materia* (Venetiis).

—— 1565. *Commentarii in VI libros Pedacii Dioscoridis Anazarbei de Medica Materia* (Venetiis).

—— 1598. *Commentarii in VI libros Pedacii Dioscorides Anazarbei de Medica Materia* (ed. Casparo Bauhino, Francofurti ad Moenum).

Maughan, W. C. 1873. *The Alps of Arabia: travels in Egypt, Sinai, Arabia and the Holy Land* (London).

Maunsell, F. R. 1896. "East Turkey in Asia and Armenia." *Scottish Geographical Magazine* 12: pp. 225–241.

Maurizio, D. A. 1932. *Histoire de l'alimentation végétale depuis la préhistoire jusqu'a nos jours* [1st ed. 1916–1917] (trad. F. Gidon, Paris).

Mauro [Camaldolese]. 1806. *Il Mappamondo di Fra Mauro Camaldolese* (ed. P. Zurla, Venezia).

McCoy, F. 1885–1890. *Natural History of Victoria* (2 v., Melbourne).

Mekhithar, Heratsi. 1908. *Mechithar: Trost bei Fiebern* (ed. and trans. E. Seidel, Leipzig).

Mela, Pomponius. 1967. *De Chorographia Libri Tres* (ed. C. Frick, Stuttgart).

Mendeville, John. 1953. *Travels* (ed. and trans. M. Letts, Hakluyt Society, 2 v. [CI, CII], London).

Merat, F. V. and A. J. de Lens. 1829–1834. *Dictionnaire universel de matière médicale* (6 v., Paris).

Meulen, D. Van Der and H. Von Wissmann. 1932. *Ḥaḍramaut* (Leiden).

Meyendorff, G. de. 1826. *Voyage d'Orenbourg à Boukhara, fait en 1820 . . . et revu par M. le Chevalier A. Jaubert* (Paris).

Meyer, C. A. 1830. *Reise durch das soongorische Kirgisen-Steppe*. In: C. F. von Ledebour *Reise durch das Altai-Gebirge* (2 v., 1829–1830, Berlin) 2: pp. 171–522.

—— 1847. "Bericht über die sogenannte Manna von Sawel." *Bulletin de la Classe Physico-Mathématique de l'Académie Impériale des Sciences de Saint Petersbourg* 6, 15: pp. 236–239.

Meyerhof, M. 1918. "Der Bazar der Drogen und Wohlgerüche in Kairo." *Archiv für Wirtschaftsforschung im Orient* 3–4: pp. 1–40, 185–218.

—— 1931. "Alī aṭ-Ṭabarī's "Paradise of Wisdom," one of the oldest compendiums of medicine." *Isis* 15: pp. 6–54.

—— 1938. "Essai sur les noms portugais de drogues dérivés de l'arabe." *Petrus Nonius* 2: pp. 1–8.

—— 1944–1945a. "The background and origins of Arabian pharmacology." *Ciba Symposia* 6: pp. 1847–1856.

—— 1944–1945b. "Pharmacology during the golden age of Arabian medicine." *Ciba Symposia* 6: pp. 1857–1867.

—— 1947. "The earliest mention of a manniparous insect." *Isis* 37: pp. 31–36.

Michaélis, Johann David. 1774. *Recueil de questions, proposées à une société de savants, qui par ordre de sa majesté Danoise font le voyage de l'Arabie* (Amsterdam and Utrecht).

Mignan, R. 1839. *A winter journey through Russia, the Caucasian Alps, and Georgia . . . . into Koordistaun* (2 v., London).

Miller, J. I. 1969. *The spice trade of the Roman Empire* (Oxford).

Miquel (Prof.) 1846. [Die Manna . . . *Lichen esculentus* Pall.] *Botanische Zeitung* 4, 24: col. 416.

Moldenke, H. N. and A. L. Moldenke. 1952. *Plants of the Bible* (Waltham, Mass.).

Molina, Giovanni Ignazio. 1809. *The Geographical, Natural, and Civil History of Chili* [1788–1795] (trans. from the Italian, 2 v., London).

Monier-Williams, M. 1899. *A Sanskrit-English Dictionary* (Oxford).

Montagne, J. F. C. 1846. *Flore d'Algérie*. In: *Exploration scientifique de l'Algérie pendant les années 1840–1842: Botanique* (Paris) 1.

More, R. 1750. "Part of a letter from Robert More Esq. to Mr W. Watson F.R.S., concerning the method of gathering manna near Naples." *Philosophical Transactions of the Royal Society of London* 46: pp. 470–471.

Morgan, J. de. 1894–1905. *Mission scientifique en Perse* (5 v., Paris).

Morier, J. 1812. *A journey through Persia, Armenia and Asia Minor to Constantinople* (London).

Moringlane (M.), (M.) Duponchel, and (M.) Bonastre. 1822. "Manne de Briançon." *Journal de Pharmacie* 8: p. 335.

Morison, Antoine. 1704. *Relation historique d'un voyage nouvellement fait au Mont de Sinai et à Jerusalem* [1697] (Toulon).

Mounsey, A. H. 1872. *A journey through the Caucasus and the interior of Persia* (London).

Mountford, C. P. 1945. "Australian natives eat "live sugar"." *Natural History* 54: pp. 158–159.

139

Mudie, R. 1829. *The picture of Australia* (London).

—— 1832–1833. "Observations on some medicinal products of Australian plants." *Transactions of the Medico-Botanical Society of London* 1832–1833: pp. 23–25.

Müller, C. (ed.) 1846. *Scriptores Rerum Alexandri Magni* (Parisiis).

Müller, F. Von. 1884. *Eucalyptographia* (Melbourne) Decade X.

Müller, J. 1858. "Ueber *Chlorangium Jussufii* H. F. Link." *Botanische Zeitung* 16, 14: pp. 89–90.

Munby, G. 1847. *Flore de l'Algérie* (Paris).

—— 1850. "On the vegetable productions of Algiers." *Report of the 19th meeting of the British Association for the Advancement of Science* 1850: p. 71.

—— 1866. *Catalogus plantarum in Algeria sponte nascentium* (2nd ed., London).

Mundy, G. C. 1855. *Our Antipodes* (London).

Mundy, Peter. 1905–1936. *The Travels of Peter Mundy in Europe and Asia, 1608–1667* (ed. R. C. Temple, Hakluyt Society, 6 v. [XVII, XXXV, XLV, XLVI, LV, LXXVIII], London).

Murav'ev, N. 1871. *Journey to Khiva through the Turkoman country* (Calcutta).

Murchison, R. I. 1864. "On a recent fall of manna in Asia Minor." *The Reader* 1864: pp. 173–174.

Murray, G. W. 1956. "Felix Fabri's pilgrimage from Gaza to Mount Sinai and Cairo, A.D. 1483." *Geographical Journal* 122, 3: pp. 335–342.

Murray, Johann Anders. 1776–1792. *Apparatus Medicaminum* (6 v., Goettingae).

Muschler, R. 1912. *A Manual Flora of Egypt* (2 v., Berlin).

Musil, A. 1927. *Arabia Deserta* (New York).

Musset, C. 1879. "Observations sur une pluie de séve." *Comptes Rendus Hebdomadaires des Séances de l'Académie des Sciences* (Paris) 88: pp. 306–307.

Muwaffiq ibn 'Alī, Abū Mansūr. 1968. *Die pharmakologischen Grundsätz* [Liber fundamentorum pharmacologiae] *des Abu Mansur Muwaffak bin Ali Harawi* (ed. and trans. Abdul-Chalig Achundow [Kobert's Historische Studien aus dem Pharmakologischen Institut der Universität Dorpat, 3, 1893] reprinted In: *Historische Studien zur Pharmakologie der Greichen, Römer und Araber*, Leipzig), pp. 137–414.

Myers, J. G. 1929. *Insect singers: a natural history of the Cicadas* (London).

Nábělek (Fr.) 1929. *Inter Turcico-Persicum IV.* Publications de la Faculté des Sciences de l'Université Masaryk (Brno).

Nassir, Abou Mouyn ed Din [fils de Khosrau]. 1881. *Sefer Namèh: relation du voyage de Nassiri Khosrau en Syrie, en Palestine, en Égypte, en Arabie et en Perse* (437–444 [1035–1042]) (ed. and trans. C. Schefer, Paris).

Naville, E. 1894. *The temple of Deir el-Bahari* (London).

Niccolò [of Poggibonsi] (Fra). 1945. *A Voyage beyond the Seas* (ed. and trans. T. Bellorini and E. Hoade, Studium Biblicum Franciscanum, Jerusalem).

Nicolaisen, J. 1963. *Ecology and culture of the pastoral Tuareg* (Copenhagen).

Niebuhr, Karsten. 1773. *Description d'Arabie* (Copenhagen).

—— 1972. *Travels through Arabia and other countries in the East* [1761–1764] (trans. R. Heron, 2 v., Edinburgh and Dublin).

Niedenzu, F. 1895. *De Genere Tamarice* (Brunsbergae).

Niessl, G. Von. 1865. "Ueber die bei Charput gefallene Manna." *Verhandlungen des Naturforschenden Vereines in Brünn* (Brno) 3: pp. 74–75.

Noe, Bianco. 1598. *Viaggio da Venetia al Santo Sepolcro et al Monte Sinai* (Venetia).

Noetling, F. 1911. "The food of the Tasmanian aboriginees." *Papers and Proceedings of the Royal Society of Tasmania* 63: pp. 279–305.

Nylander, W. 1854. "Études sur les lichens de l'Algérie." *Mémoires de la Société Impériale des Sciences Naturelles de Cherbourg* 2: pp. 305–344.

—— 1857. "Prodromus lichenographie Galliae et Algeriae." *Actes de la Société linnéenne de Bordeaux* 3iéme sér. 1: pp. 249–467.

—— 1858. "Enumération générale des lichens, avec l'indication sommaire de leur distribution géographique." *Mémoires de la Société Impériale des Sciences Naturelles de Cherbourg* 5: pp. 85–146.

—— 1878. "Symbolae quaedam ad lichenographiam Sahariensem." *Flora* 36: pp. 337–345.

Ocampo, Melchor. 1900–1901. *Obras completas* (México).

Odoric [of Pordenone] (Fr.) 1891. *Les voyages en Asie au XIV e. siècle du Frère Odoric de Pordenone* (ed. H. Cordier, Paris).

—— 1904. *Incipit Itinerarium fratris Odorici fratrum minorum de mirabilibus Orientalium Tartarorum.* In: Richard Hakluyt *The Principal Navigations, Voyages, Traffiques and Discoveries of the English Nation* (12 v., 1903–1905, Glasgow) 4: pp. 371–444.

—— 1913. *The Travels of Friar Odoric.* In: *Cathay and the Way Thither: being a collection of medieval notices of China* (ed. and trans. H. Yule, revised by H. Cordier, Hakluyt Society, 2nd ser. 4 v. [XXXIII, XXXVII, XXXVIII, XLI], London) 2.

Oliver, D., W. T. Thiselton-Dyer, D. Prain, and A. W. Hill (eds.) 1868–1937. *Flora of Tropical Africa* (10 v., London).

Oliver, S. P. 1886. *Madagascar* (2 v., London).

Oliver, Guillaume Antoine. 1801–1807. *Voyage dans l'Empire Othoman, l'Égypte et la Perse* (3 v., Paris).

Oribasius. 1851–1876. *Ouevres d'Oribase* (ed. C. Daremberg and U. C. Bussemaker, 6 v., Paris).

—— 1940. *Oribasius Latinus* (ed. H. Morland, Symbolae Osloenses, fasc. suppl. X, Osloae).

O'Rorke, (M.) 1860. "La manne des Hébreux." *Journal de pharmacie et de chimie* 3ième sér. 37: pp. 412–419.

Orta, Garcia Da. 1576. *Due libri dell' Historia de i Semplici, Aromati, et altre cose, che vengono portate dall' Indie Orientali* (ed. and trans. Annibale Briganti, Venetia).

—— 1913. *Colloquies on the Simples and Drugs of India* [1563] (ed. and trans. C. R. Markham, London).

O'Shaughnessy, W. B. 1842. *Bengal Dispensatory* (London).

Osten-Sacken (Fr. v. d.) and F. J. Ruprecht. 1869. "Sertum Tianschanicum: Botanische Ergebnisse einer Reise im mittleren Tian-Schan." *Mémoires de l'Académie Impériale des Sciences de Saint-Pétersbourg* 7ième sér. 14, 4: pp. 1–74.

Otter, Jean. 1748. *Voyage en Turquie et en Perse* (2 v., Paris).

Ouseley, W. 1819–1825. *Travels in various countries in the East, more particularly in Persia* (3 v., London).

Ovid. 1966–1968. *Metamorphoses* (ed. and trans. F. J. Miller, 2 v., Loeb Classics London).

Ovington, J. 1929. *A voyage to Surat in the year 1689* (ed. H. G. Rawlinson, London).

Owen, W. F. W. 1833. *Narrative of voyages to explore the shores of Africa, Arabia, and Madagascar* (2 v., London).

Ozenda, P. 1977. *Flore de Sahara* (2nd ed., Paris).

Palea, Angelus (Fr.) and Fr. Bartholomaeus ab Urbe veterum. 1550. *Antidotarium Joannis Filii Mesuae Censura* (Lugduni).

Pallas, Peter Simon, 1771–1776. *Reise durch verschiedene Provinzen des Russischen Reiches* (3 v., St. Petersbourg).

—— 1784–1815. *Flora Rossica* (2 v., Petropoli and Berlin).

—— 1788–1793. *Voyages en différentes provinces de l'empire de Russie, et dans l'Asie septentrionale* [1768–1774] (trad. M. Gauthier de la Peyronie, 5 v., Paris).

Palmer, E. 1884. "On plants used by the natives of north Queensland, Flinders and Mitchell rivers, for food, medicine …" *Journal and Proceedings of the Royal Society of New South Wales* 17: pp. 93–113.

Palmer, E. H. 1871. *The Desert of the Exodus* (Cambridge).

Palmer, H. S. 1892. *Sinai from the Fourth Egyptian Dynasty to the present day* (London).

—— 1906. *Sinai* (London).

Pampanini, R. 1914. *Plantae Tripolitanae* (Firenze).

Paris, R. and G. Dillemmann. 1960. *Medicinal Plants of the Arid Zones* (Paris).

Parrot, F. 1834. *Reise zum Ararat* (2 v., Berlin).

Pasi, Bartolommeo di. 1521. *Tariffa di pesi e mesure correspondenti dal Levante al Ponente* (Vinetia).

—— 1540. *Tariffa de i pesi, e misure corrispondenti dal Leuante al Ponente* (Vinetia).

Passmore, F. W. 1891. "On the carbohydrates of manna from *Eucalyptus gunnii* Hook., and of eucalyptus honey." *Pharmaceutical Journal and Transactions* 21: pp. 717–720.

Paterson, W. 1900. [Letter to Sir Joseph Banks]. In: *Historical Records of New South Wales* (ed. A. Britton and F. M. Bladen, 7 v., 1892–1901, Sidney) 6: p. 768.

Paulsen, O. 1912. *Studies on the vegetation of the Transcaspian Lowlands* (Copenhagen).

Paulus [Aegineta] 1844–1847. *The Seven Books of Paulus Aegineta* (ed. and trans. F. Adams, 3 v., London).

—— 1914. *Paulos von Aegina* (ed. and trans. I. Berendes, Leiden).

Pegolotti, Francesco Balducci. 1936. *La Practica della Mercatura* [ca. 1310–1340] (ed. A. Evans, Cambridge, Mass.).

Pemel, Robert. 1652–1653. *A treatise of the nature and qualities of such simples as are most frequently found in medicines* (London).

Pesman, M. W. 1962. *Flora Mexicana* (Globe, Arizona).

Pétis de la Croix, François. 1722. *History of Genghiscan the Great* (London).

141

Petrie, W. M. Flinders. 1906. *Researches in Sinai* (London).

Philby, H. St. J. B. 1922. *The heart of Arabia* (2 v., London).

Phipson, T. L. 1856–1857. "On the production of mannite by marine plants." *Pharmaceutical Journal and Transactions* 16: p. 530.

—— 1859. "Production de la mannite par les algues marines." *Journal de Pharmacie et de Chimie* 3ième sér. 35: p. 314.

Pickering, C. 1854–1876. *The Geographical Distribution of Plants and Animals* (2 v., Boston and London).

—— 1879. *Chronological History of Plants* (Boston).

Pickhall, M. (trans.) 1938. *The meaning of the Glorous Qur'ân* (2 v., Hyderabad-Deccan).

Picolo, Francesco Maria. 1715. *Mémoire touchant la Californie.* In: J. F. Bernard (ed.) *Recueil de voyages au Nord* (8 v., 1715–1727, Amsterdam) 3: pp. 278–300.

Pierret, P. 1875. *Vocabulaire hiéroglyphique* (Paris).

Pilter, W. T. 1917. "The manna of the Israelites." *Proceedings of the Society of Biblical Archaeology* 39: pp. 155–167, 187–206.

Pinkerton, J. 1811. *A copious and circumstantial description of the great empire of Persia* [after Herbert, Chardin, Tavernier . . .] In: *A general collection of the best and most interesting voyages and travels in all parts of the world* (17 v., 1808–1814, London) 9: pp. 168–198.

Pitard, C. J. and L. Corbière. 1913. *Exploration scientifique du Maroc,* premier fascicule [botanique] (Paris).

Pitra, A. 1868. "Ueber *Chlorangium esculentum.*" *Hedwigia* 7, 1: pp. 7–12.

Planchon, F. G. and E. Collin. 1895–1896. *Les drogues simples d'origine végétale* (2 v., Paris).

Platearius, Johannes [Mattheus]. 1524. *Liber de Simplici Medicina* (Lugduni).

—— 1913. *Le livre des simples médicines: traduction française du Circa instans de Platearius* (ed. et trad. P. Dorveaux, Société française d'histoire de la médicine, Paris).

—— ca. 1972. *Een Middelnederlandse Versie van de Circa Instans van Platearius* (ed. L. J. Vandewiele, Oudenarde).

Pliny [The Elder]. 1961–1968. *Natural History* (ed. and trans. H. Rackham, W. H. S. Jones and D. E. Eichholz, Loeb Classics, 10 v., London).

Pococke, Richard. 1737. *A Description of the East* (2 v., London).

Polak, J. E. 1865. *Persien, das Land und seine Bewohner* (2 v., Leipzig).

Polyaenus. 1793. *Polyaenus's Stratagems of War* (ed. and trans. R. Shepherd, London).

Pomet, Pierre. 1694. *Histoire générale des drogues* (2 v., Paris).

—— 1709. *Droguier curieux: catalogue des drogues, simples et composées* (2nd ed., Paris).

—— 1725. *A compleat history of drugs* (2nd ed. of the translation, London).

Pontanus, Johannes Jovianus. 1513. *Pontani Opera Poetica* (Venetiis).

Porter, R. K. 1821–1822. *Travels in Georgia, Persia, Armenia and Ancient Babylonia* [1817–1820] (2 v., London).

Post, G. E. 1932–1933. *Flora of Syria, Palestine and Sinai* (2 v., Beirut).

Pottinger, H. 1816. *Travels in Beloochistan and Sinde* (London).

Prescott, H. F. M. 1957. *Once to Sinai: the further pilgrimage of Friar Felix Fabri* (London).

Price, W. 1825. *Journal of travels of the British Embassy to Persia* (London).

Proust, (M.) 1806. "Mémoire sur le sucre de raison." *Annales de Chimie* 57: pp. 131–174.

Ptolemaeus, Claudius, 1540. *Geographia Universalis, Vetus et Nova Complectens* (Basileae).

Quer, Joseph. 1762–1784. *Flora Española* (6 v., Madrid).

Rabino, H. L. 1911. *Report on Kurdistan* (Simla).

—— 1916. *Les Tribus du Louristan* (Paris).

Raby, (M.) 1889. "Sur deux matières sucrées retirées des mannes Chirkésht et Bidenguébine." *Union Pharmaceutique* 1889: pp. 201–208.

Radcliffe, William. 1789. *The natural history of East Tartary* (London).

Ranke, Leopold Von. 1908. *The History of the Popes* [*Die römischen Päpste im XVI und XVII Jahrhundert,* 3 v., 1838–1839] (3 v., London).

Raumer, Karl Von. 1837. *Der Zug der Israeliten aus Aegypten nach Canaan* (Leipzig).

Rauwolf, Leonhard. 1693. *Travels into the Eastern Countries* [1573–1576]. In: John Ray (ed.) *A collection of curious travels and voyages* (2 v., London).

—— 1755. *Flora orientalis* [1573–1575] (published by J. F. Gronovius, Lugduni Batavorum).

Rawlinson, (Major). 1839. "Notes on a march from Zoháb, at the foot of the Zagros, along the mountains to Khúzistán (Susiana), and from thence through the province of Luristan to Kirmánsháh, in the year 1836." *Journal of the Royal Geographical Society* 9: pp. 26–116.

Ray, John (ed.) 1693. *A collection of curious travels and voyages* [Appendix: *Stirpium Orientalium Rariorum Catalogus*] (2 v., London).

—— 1848. *Correspondence of John Ray* (ed. E. Lankester, Ray Society, London).

Read, B. E. 1936. *Chinese medicinal plants from the Pen Ts'ao Kang Mu* (1956) [of Li Shih-chên] (3rd ed., Peking).

Read, B. E. and Lui Ju-Ch'iang. 1927. *Plantae Medicinalis Sinensis* (Peking).

Reaumur, R. A. F. de. 1734–1742. *Mémoires pour servir à l'histoire des insectes* (6 v., Paris).

Rechinger, K. H. (gen. ed.) 1963-. *Flora Iranica* (fascules, Graz).

Regnault, J. 1902. *Médicine et pharmacie chez les Chinois et chez les Annamites* (Paris).

Reichardt, H. W. 1864. "Ueber die Manna-Flechte, *Sphaerothallia esculenta* Nees." *Verhandlungen Zoologisch-botanischen Gesellschaft in Wien* 14: pp. 553–560.

Reischauer, E. O. 1955. *Ennin's Travels in T'ang China* (New York).

Reissek, S. 1847. "Ueber die Natur des kürzlich in Klein-Asien vom Himmel gefallenen manna." *Berichte über die Mittheilungen von Freunden der Naturwissenschaften in Wien* 1: pp. 195–201.

Renard, E. and E. Lacour. 1880. "De la manne du désert, ou manne des Hébreux." *Journal de médicine et de pharmacie de l'Algérie* 1880: pp. 3–20.

Renaud, H. P. J. 1935. "La contribution des Arabes à la connaissance des espèces végétales: les botanistes Musulmans." *Bulletin de la Société des Sciences Naturelles du Maroc* 15: pp. 58–71.

Renaud, H. P. R. and G. S. Colin (ed. et trad.) 1934. *Tuḥfat al-aḥbāb: glossaire de la matière médicale marocaine* (Paris).

Renodaeus, Johannes. 1609. *Institutionum Pharmaceuticarum* (Paris).

Rhind, W. 1855. *A History of the Vegetable Kingdom* (London).

Ricettario Fiorentino. 1548. *Recettario Utilissimo et Molto Necessario* [Collegio de' Medici, Firenze] (In Venetia).

—— 1550. *El Ricettario del l'Arte et Universita de Medici, et Spetiali della Citta di Firenze* (Fiorenza).

—— 1567. *Ricettario Fiorentino* (Firenze).

—— 1597. *Ricettario Fiorentino, di Nuovo Illustrato* (Fiorenza).

—— 1968. *Nuouo receptario composto dal famossisimo Chollegio degli eximii doctori della arte et medicina della inclita cipta di Firenze* (fasc. of ed. of 1498, Firenze).

Rich, C. J. 1836. *Narrative of a residence in Koordistan* (2 v., London).

Riedesel, J. H. Von. 1773. *Travels through Sicily and that part of Italy formerly called Magna Graecia* (trans. J. R. Forster, London).

Riegler, L. 1852. *Die Turkei und deren Bewohner* (2 v., Wien).

Ritter, K. 1822–1859. *Die Erdkunde* (19 v., Berlin).

—— 1866. *The comparative geography of Palestine and the Sinaitic peninsula* (trans. W. L. Gage, 4 v., Edinburgh).

Robinson, E. and E. Smith. 1841. *Biblical Researches* (3 v., London).

Robinson, Tancred. 1717. "Miscellaneous observations made about Rome, Naples and some other countries, in the years 1683 and 1684." *Philosophical Transactions of the Royal Society of London* 29: pp. 473–483.

Rockhill, W. W. 1915. "Notes on the relations and trade of China with the eastern archipelago and the coast of the Indian Ocean during the 14th century." *T'oung Pao* 16: pp. 604–626.

Rohlfs, G. F. 1875. *Expedition zur Erforschung der libyschen Wüste* (Cassel).

Rolt, Richard. 1761. *A New Dictionary of Trade and Commerce* (London).

Rosenmüller, E. F. C. 1840. *The mineralogy and botany of the Bible* (trans. T. G. Repp and N. Morren, Edinburgh).

Roth, W. E. 1901. "Food: its search, capture and preparation." *North Queensland Ethnography, Bulletin* 3: pp. 7–31.

Rousseau, J. J. 1794. *Letters on the Elements of Botany* (trans. Thomas Martyn, 4th ed., London).

Roxburgh, W. 1820–1832. *Flora Indica* (3 v., Serampore and London).

Royle, J. F. 1837. *An essay on the antiquity of Hindoo medicine* (London).

—— 1839. *Illustrations of the botany and other branches of the natural history of the Himalayan mountains and of the flora of Cashmere* (2 v., London).

—— 1840. *Essay on the productive resources of India* (London).

—— 1876. *A manual of materia medica* (6th ed., London).

Rüppell, E. 1829. *Reisen in Nubien, Kordofan und dem peträischen Arabien* (Frankfurt).

Russell, A. 1794. *The Natural History of Aleppo* (2 v., London).

S., F. 1938. "La manne de Madagascar." *L'Agronomie Coloniale* no. 242: p. 54.

Sabeti, H. 1966. *Native and exotic trees and shrubs of Iran* (Tehrān).

Sahagún, Bernardino de. 1963. *Florentine Codex: General History of the Things of New Spain* [ca. 1570] (ed. and trans. C. E. Dibble and A. J. O. Anderson, book XI, Sante Fé).

143

Saladinus, Asculanus. 1488. *Compendium Aromatariorum* (Bologna).

—— 1581. *Compendium Aromatariorum*. In: Ioannis Mesuae [Ibn Māsawaih] *Medici Clarissimi Opera* (Venetiis), pp. 251–264.

—— 1953. *Compendium Aromatariorum* (ed. S. Muntner, Tel Aviv).

Salmasius, Claudius. 1689. *Plinianae Exercitationes in Caji Julii Solini Polyhistora* (2 v., Trajecti ad Rhenum).

Salmon, G. 1906. *Sur quelques noms de plantes en arabe et en berbère*. Archives Marocaines 8 (Paris).

Santamaria, F. J. 1942. *Diccionario General de Americanismos* (3 v., Madrid).

Sarton, G. 1927–1948. *Introduction to the History of Science* (3 v. in 5, Baltimore).

Sassoon, G. and R. Dale. 1976. "Deus est machina?" *New Scientist* 70: pp. 22–24.

Savary des Bruslons, J. and P. Savary des Bruslons. 1742. *Dictionnaire universel de commerce* [1723] (3 v., Genève).

—— 1751–1755. *The Universal Dictionary of Trade and Commerce* [1723] (trans. with additions, M. Postlethwayt, 2 v., London).

Scaliger, Julius Caesar. 1557. *Exotericarum Exercitationem Liber Quintus Decimus de Subtilitate, ad Hieronymum Cardanum* (Lutetiae).

Schafer, E. H. 1977. "T'ang." In: K. C. Chang (ed.) *Food in Chinese Culture: anthropological and historical perspectives* (New Haven and London), pp. 85–140.

Schlimmer, J. 1874. *Terminologie Medico-Pharmaceutique et Anthropologique Française-Persane* (Theheran).

Schoenherr, C. J. 1833–1838. *Genera et Species Curculionidum* (5 v., Paris).

Schoff, W. H. (ed. and trans.), 1912. *The Periplus of the Erythraean Sea. Travel and Trade in the Indian Ocean by a Merchant of the 1st Century* (London).

Schott, W. 1876. "Zur Uigurenfrage." *Philologische und Historische Abhandlungen der Königlichen Akademie der Wissenschaften zu Berlin* 1876: pp. 27–57.

Schubert, G. H. Von. 1838–1839. *Reise in das Morgenland* (3 v., Erlangen).

Schumacher, C. F. 1801. *Enumeratio Plantarum in Partibus Saellandiae Septentrionalis et Orientalis* (2 v., Hafniae).

Schwarz, O. 1936. "Entwurf zu einem natürlichen System der Cupuliferen und der Gattung Quercus L." *Notizblatt des Botanischen Gartens und Museums zu Berlin-Dahlem* 13: pp. 1–22.

Schwarz, P. 1896–1936. *Iran im Mittelalter nach den arabischen Geographen* (9 v., Leipzig).

Schweinfurth, G. 1912. *Arabische Pflanzennamen aus Aegypten, Algerien und Jemen* (Berlin).

—— 1918. "Über Brotbacken mit Zusatz von Flechten in Ägypten." *Archiv für Wirtschaftsforschung im Orient* 3, 3–4: pp. 439–443.

Sébire, A. 1899. *Les plantes utiles du Sénégal* (Paris).

Seemann, B. 1864. "On the recent fall of manna in Asia Minor." *The Reader* 1864: p. 205.

Seetzen, U. J. 1808. "Auszug aus einem Schreiben des Russisch-Kaiserlichen Kammer-Assessors." *Monatliche Correspondenz zur Beförderung der Erd- und Himmels-Kunde* 17: pp. 132–163.

—— 1854–1859. *Reisen durch Syrien, Palästina, Phönicien, die Transjordan-Länder, Arabia Petraea und Unter-Aegypten* (4 v., Berlin).

Segneri, Paolo. 1879. *The Manna of the Soul* [Manna Animae] (4 v., London).

Seneca, Lucius Annaeus. 1962–1967. *Epistulae Morales* (ed. and trans. R. M. Gummere, Loeb Classics, 3 v., London).

Sestini, Domenico. 1788. *Descrizione di vari prodotti dell' Isola di Sicilia* (Firenze).

Shantz, H. L. and C. F. Marbut. 1923. *The Vegetation and Soils of Africa* (American Geographical Society, Research Series 13, New York).

Shaw, Thomas. 1738. *Travels or observations relating to several parts of Barbary and the Levant* (Oxford).

Sibthorp, John. 1806–1813. *Florae Graecae Prodromus* (2 v., Londini).

—— 1806–1840. *Flora Graeca* (10 v., Londini).

Sickenberger, E. 1890. *Les plantes égyptiennes d'Ibn al-Beithar*. Bulletin de l'Institut Egyptien, 2ième sér. 10 (Le Caire).

—— 1893. *Die einfachen Arzneistoffe der Araber im 13 Jahrhunderte christl. Zeitr. von E. Sickenberger*. Pharmaceutische Post 1891–1893, als Sonderabdruck erschienen (Wien).

Simmonds, P. L. 1895. "Notes on some saps and secretions used in pharmacy." *The American Journal of Pharmacy* 67: pp. 128–135.

Singh, Thakur Balwant, and K. C. Chunekar. 1972. *Glossary of Vegetable Drugs in Bṛhattrayī*. Chowkhamba Sanskrit Studies 87 (Varanasi).

Smith, A. L. 1921. *Lichens* (Cambridge).

Smith, F. P. 1871. *Contributions towards the Materia Medica of China* (Shanghai and London).

Smith, H. G. 1897. "On the saccharine and astringent exudations of the Grey Gum, *Eucalyptus punctata* D.C." *Journal of the Royal Society of New South Wales* 31: pp. 177–194.

Smith, J. 1882. *A dictionary of popular names of the plants which furnish the natural and acquired wants of man* (London).

Smith, R. G. 1902. "A gum (levan) bacterium from a saccharine exudate of *Eucalyptus stuartiana*." *Proceedings of the Linnean Society of New South Wales* 27: pp. 230–236.

Smith, W. (ed.) 1863. *A dictionary of the Bible* (3 v., London).

Sontheimer, J. 1840–1842. *Grosse Zusammenstellung über die Kräfte der bekannten einfachen Heil- und Nahrungsmittel von Ebn Baithar* (2 v., Stuttgart).

Soubeiran, E. 1840. *Nouveau traité de pharmacie théoretique et pratique* (2 v., Paris).

Spencer, W. B. and F. J. Gillen. 1938. *Native Tribes of Central Australia* (London).

Sprengel, C. P. J. 1807–1808. *Historia rei herboriae* (2 v., Amsteldami).

—— 1825–1828. *Carolus Linnaeus: Systema Vegetabilium* (5 v., Gottingae).

Stahl, A. F. Von. 1924. "Notes on the march of Alexander the Great from Ecbatana to Hyrcania." *Geographical Journal* 64: pp. 312–329.

Stanley, A. P. 1856. *Sinai and Palestine, in connection with their history* (London).

Steiner, J. 1898. "Prodromus einer Flechtenflora des griechischen Festlandes." *Sitzungsberichte der Mathematisch-Naturwissenschaftlichen Classe der Kaiserlichen Akademie der Wissenschaften in Wien* 107: pp. 103–189.

—— 1899. "Flechten aus Armenien und dem Kaukasus." *Österreichische botanische Zeitschrift* 49: pp. 248–254, 292–295.

—— 1921. "Lichenes aus Mesopotamien und Kurdistan sowie Syrien und Prinkipo." *Annalen des Naturhistorischen Museums in Wien* 34: pp. 1–68.

Steingass, F. J. 1957. *Persian-English Dictionary* (London).

Stephens, J. L. 1838. *Incidents of travel in Egypt, Arabia Petraea, and the Holy Land* (2 v., London).

Steven, Christian Von. 1811. *Catalogue des plantes rares ou nouvelles, observées pendant un voyage autour du Caucase oriental* (St. Petersburg).

—— 1856–1857. "Verzlichniss der auf der taurischen Halbinsel wildwachsenden Pflanzen." *Bulletin de la Société Imperiale des Naturalistes* (Moscow) 29, 1: pp. 134–276; 29, 2: pp. 121–186, 339–418; 30, 1: pp. 325–398; 30, 2: pp. 65–160.

Stevens, John. 1715. *The History of Persia* (London).

Stillé, A. 1868. *Therapeutics and Materia Medica* (2 v., 3rd ed., Philadelphia).

Stillingfleet, B. 1762. *Miscellaneous tracts relating to natural history, husbandry and physick* (London).

Stokes, J. L. 1846. *Discoveries in Australia* [1837–1843] (2 v., London).

Strabo. 1860–1869. *The Geography of Strabo* (ed. and trans. H. L. Jones, 8 v., Loeb Classics, London).

Stuart, G. A. 1928. *Chinese Materia Medica* (Shanghai).

Sturtevant, E. L. 1972. *Edible Plants of the World* [1st ed. 1919] (ed. U. P. Hedrick, New York).

Suárez y Núñez, Miguel Gerónimo. 1778–1791. *Memorias instructivas y curiosas sobre agricultura, commercio* . . . . (12 v., Madrid).

Suriano, Francesco. 1949. *Treatise on the Holy Land* (ed. and trans. T. Bellorini and E. Hoade, Studium Biblicum Franciscanum, Jerusalem).

Swann, A. T. 1910. *Fighting the slave-hunters in Central Africa* (London).

Swinburne, Henry. 1783–1785. *Travels in the Two Sicilies* (2 v., London).

Sykes, P. M. 1906. "A Fifth Journey in Persia." *Geographical Journal* 28: pp. 425–453.

Syme, J. T. Boswell and J. E. Sowerby. 1872. *English Botany* (12 v., 1863–1886, London) 11.

Sylvius, Jacobus. 1548. *Methodus medicamenta componendi, ex simplicibus* (Lugduni).

Tabeeb, Meerza Jiáfer. 1819. ["On Persian manna."] *Asiatic Journal* 7: p. 268.

Takhtadzhiana, A. L. (ed.) 1954–1966. *Flora Armenii* (5 v., Erevan).

Targioni-Tozzetti, Giovanni. 1768–1779. *Relazioni d'alcuni viaggi fatti in diverse parti della Toscana* (12 v., 2nd ed., Firenze).

Tariffa. 1791. *Tariffa delle Gabelle per Firenze* (Firenze).

Tate, G. P. 1909. *The frontiers of Baluchistan* (London).

Tease, G. E. 1936. "Gums of the tragacanth type." *Pharmaceutical Journal* 137: pp. 206–208.

Teesdale, M. J. 1897. "The manna of the Israelites." *Science Gossip*, new series, 3: pp. 229–233.

Teixeira, Pedro. 1902. *The Travels of Pedro Teixeira* (ed. and trans. W. F. Sinclair, Hakluyt Society, second series, IX, London).

Temple, A. A. 1929. *Flowers and trees of Palestine* (London).

Tennent, J. Emerson. 1860. *Ceylon* (2 v., London).

Tepper, J. G. O. 1883. "Remarks on the manna or lerp insect of South Australia." *Journal of the Linnean Society (Zoology)* 17: pp. 109–111.

Thénard, (Prof.) 1828. ["Durch die Winde fortgeführter nahrhafter Stoff."] *L. F. von Froriep's Notizen aus dem Gebiete der Natur- und Heilkunde* 22, 4: col. 55.

Thenaud, Jean. 1884. *Le Voyage d'Outremer de Jean Thenaud* [1512] (ed. C. Schefer, Paris).

Theophrastus. 1961–1968. *Enquiry into Plants* (ed. and trans. A. Hort, 2 v., Loeb Classics, London).

Thetmarus (magister). 1851. *Iter ad Terram Sanctam anno 1217* (ed. T. Tobler, St. Galli et Bernae).

Thevenot, Jean de. 1949. *Indian travels of Thevenot and Careri* (ed. Surendranath Sen, New Delhi).

Thompson, Charles. 1744. *Travels through Turkey in Asia, the Holy Land, Arabia, Egypt, and other parts of the world* (3 v., Reading).

Thompson, R. Campbell. 1924. *The Assyrian Herbal* (London).

—— 1937. "Assyrian prescriptions for the head." *American Journal of Semitic Languages* 53: pp. 217–238.

—— 1949. *A dictionary of Assyrian Botany* (London).

Thomson, T. 1838. *Chemistry of Organic Bodies: Vegetables* (London).

Tischendorf, L. F. K. 1862. *Aus dem heiligen Lande* (Leipzig).

Tizenhauz, (M.) 1846. "Note sur une substance tombée de l'atmosphère." *Comptes Rendus Hebdomadaires des Séances de l'Académie des Sciences* (Paris) 23: pp. 452–454.

Tobler, F. 1925. *Biologie der Flechten* (Berlin).

—— 1934. *Die Flechten* (Jena).

Tomin, M. P. 1926. *Über die Bodenflechten aus den Halbwüsten von Sud-Ost-Russland* (Woronesch).

—— 1956. *List of lichens of the European part of the U.S.S.R., including the extreme north and the Crimea* [in Russian] (Minsk).

Tournefort, Joseph Pitton de. 1703. *Corollarium Institutionum Rei Herbariae* (Paris).

—— 1708. *Materia Medica* (London).

—— 1741. *A Voyage into the Levant* [trans. of *Relation d'un voyage du Levant*, Lyon, 1717] (3 v., London).

Tozer, H. F. 1881. *Turkish Armenia and Eastern Asia Minor* (London).

Treviranus, L. C. 1816. "Observationes circa plantas Orientis, cum descriptionibus novarum aliquot specierum." *Magazin der Gesellschaft Naturforschender Freunde zu Berlin* 7: pp. 145–156.

—— 1848. "Noch einiges über *Lichen esculentus* Pall." *Botanische Zeitung* 6, 52: cols. 891–894.

Trotter, Alessandro. 1915. *Flora economica della Libia* (Roma).

Tschihatchcheff, Pierre de [Petr Aleksandrovich Chikhachev] 1853–1869. *L'Asie Mineure* (8 v., Paris).

Usher, G. 1974. *A dictionary of plants used by man* (London).

Vahl, M. 1790–1794. *Symbolae botanicae* (3 v., Hauniae).

Vambéry, A. 1868. *Sketches of Central Asia* (London).

Viennot, O. 1954. *Le culte de l'arbre dans l'Inde ancienne. Textes et monuments brāhmaniques et bouddhiques* (Paris).

Villars, Dominique. 1786–1789. *Histoire des plantes de Dauphiné* (4 v., Paris).

Villiers, A. 1877. "Recherches sur le mélézitose." *Comptes Rendus Hebdomadaires des Séances de l'Académie des Sciences* (Paris) 84: pp. 35–38.

Vincentius [Bellovacensis]. 1494. *Speculum Naturale* (Venetia).

Virey, J.-J. 1818. "Recherches historiques et bibliques sur la manne des Hébreux et les mannes diverses de l'Orient." *Journal de Pharmacie* 4: pp. 120–126.

—— 1832. "Manne de l'Australie." *Journal de Pharmacie* 18: pp. 705–706.

Virigil. 1967–1969. *Eclogues, Georgics, Aeneid* (ed. and trans. R. Rushton Fairclough, 2 v., Loeb Classics, London).

Visiani, Roberto de. 1836. *Plantae quaedam Aegypti ac Nubiae* (Patavii).

—— 1864–1865. "Relazioni di una pioggia di sostanza vegetabile alimentare caduta in Mesopotamia nel marzo 1864." *Atti della reg. Istituto Veneto di scienze, lettere ed arti* 3 seri 10: pp. 284–306.

—— 1867. "Bericht über einen Regen einer vegetabilischen Nahrungs-Substanz, welcher im März 1864 in Mesopotamien niedergefallen ist." *Flora* 25: pp. 197–205, 225–230.

Vogel, V. J. 1976. "American Indian foods used as medicine." In: W. D. Hand (ed.) *American Folk Medicine* (Berkeley and Los Angeles), pp. 125–141.

Volcmann, Gottlob Israel. 1725. *Dissertatio Inauguralis Medica de Manna* (Halae Magdeburgicae).

Volkens, G. 1887. *Die Flora der Aegyptisch-Arabischen Wüste* (Berlin).

Vullers, J. A. 1855–1867. *Lexicon Persico-Latinum Etymologicum* (2 v., and suppl., Bonnae ad Rhenum).

Walther, Michael. 1633. *Tractatu de Mannâ* (Lugduni Batavorum).

Walther, Paul. 1892. *Itinerarum in Terram Sanctam et ad Sanctam Catharinam* [1482–1483] (ed. M. Sollweck, Tübingen).

Walker, W. 1958. *All the plants of the Bible* (London).

Walpers, G. 1851. "Notiz über *Lichen esculentum* Pall." *Botanische Zeitung* 9, 17: pp. 317–318.

Ward, J. S. 1893. "Note on manna collection in Sicily." *Pharmaceutical Journal and Transactions* 3rd ser., 24: pp. 381–391.

Waring, E. J. 1874. *Remarks on the use of some bazaar medicines and common medicinal plants of India* (2nd ed., London).

Watt, G. 1889–1893. *Dictionary of the Economic Products of India* (6 v. in 9, London).

Webb, P. B. 1838. *Iter Hispaniense, or a synopsis of plants collected in the southern provinces of Spain and in Portugal* (Paris and London).

——— 1841. "*Tamarix gallica* of Linnaeus." *W. J. Hooker's Journal of Botany* 3: pp. 422–431.

Weckero, Ioan Jacobo. 1617. *Antidotarium generale et speciale* (Basilea).

Wehmer, Karl. 1929. *Die Pflanzenstoffe* (Jena).

Wellsted, J. R. 1838. *Travels in Arabia* (2 v., London).

Wenrich, J. G. 1845. *Rerum ab Arabibus in Italia insulisque adjacentibus, Sicilia maxime, Sardinia, atque Corsica gestarum commentarii* (Lipsiae).

West, T. 1858. "A brief description of a singular insect production found in some parts of Australia." *Sydney Magazine of Science and Art* 1: p. 75.

Westgarth, W. 1848. *Australia Felix, or a historical and descriptive account of the settlement of Port Phillip, New South Wales* (Edinburgh).

Westwood (Mr). 1846. ["Reports on manna."] *Athenaeum* no. 974: p. 659.

Wheatley, P. 1959. "Geographical notes on some commodities involved in the Sung maritime trade." *Journal of the Malayan Branch of the Royal Asiatic Society* 32: pp. 1–140.

Wheler, George. 1682. *A journey into Greece* (London).

Wight, R. 1840–1852. *Icones plantarum Indias orientalis* (6 v., Madras).

Wilkes, C. 1845. *Narrative of the United States Exploring Expedition, 1838–1842* (5 v., London).

Willemet, R. 1787. *Lichénographie économique ou histoire des lichens utiles* (Académie des Sciences, Belles-Lettres et Arts. Mémoires . . . sur l'utilité des lichens . . . . partie 1, Lyons).

——— 1808. *Phytographie encyclopédique ou flore économique* (3 v., Paris).

Wills, C. J. 1883. *In the land of the lion and sun, or modern Persia* (London).

Woenig, F. 1886. *Die Pflanzen im alten Aegypten* (Leipzig).

Wood, G. B. and F. Bache, 1907. *The Dispensatory of the U.S.A.* (19th ed., Philadelphia).

Woodson, R. E. 1930. "Studies in the Apocynaceae I." *Annals of the Missouri Botanical Garden* 17, 1: pp. 2–213.

Woodville, W. 1790–1794. *Medical Botany* (3 v., and supplement, London).

Wooster, W. H. 1882. "How the lerp crystal palace is built." *Journal of the Microscopical Society of Victoria* 1, 4: pp. 91–94.

Wright, A. H. 1847a. ["On manna."] In: O. P. Hubbard "Notices of Kurdistan." *American Journal of Science and Arts* 2nd ser. 3: pp. 350–351.

——— 1847b. "On the manna of the scriptures." *Edinburgh New Philosophical Journal* 43: pp. 176–178.

Wulff, H. E. 1966. *The traditional crafts of Persia* (Cambridge, Mass.).

Zallones, M. P. 1809. *Voyage à Tine, l'une des îles de l'archipel de la Grèce* (Paris).

Zaman, M. 1951. "The regions of drybelts in Afghanistan." *Afghanistan* 6, 2: pp. 63–68.

Zawacki, Teodor. 1891. *Memoriale Oekonomicum de l'année 1616* (published by J. Rostafinski, Kracowie).

Zenner, J. R. 1899. "*Man hu*, Exodus 16. 15." *Zeitschrift für katholische Theologie* 23: pp. 164–166.

Zénob de Glag. 1867. *Histoire de Daron.* In: V. Langlois (ed.) *Collection des historiens anciens et modernes de l'Arménie* (2 v., 1867–1869, Paris) 1: pp. 337–355.

Zohary, M. 1950. *The Flora of Iraq* (Baghdad).

—— 1961. "Oak species of the Middle East." *Bulletin of the Research Council of Israel* 9, 4: pp. 161–186.

—— 1973. *Geobotanical Foundations of the Middle East* (2 v., Stuttgart and Amsterdam).

Zopf, W. 1896. "Uber den Nutzen der Flechten." *Die Natur* (Halle) 45, 16: pp. 185–187.

Zorzi, Alessandro. 1958. *Ethiopian Itineraries, circa 1400–1524, including those collected by Alessandro Zorzi at Venice in the years 1519–1524* (ed. O. G. S. Crawford, trans. R. Radford, Hakluyt Society, 2nd. ser., CIX, London).

INDEX GENERUM ET SPECIERUM

INDEX NOMINUM

153

INDEX LOCORUM

159